14.58

DISCOVERING
CALCULUS
WITH
MATHEMATICA™

BART BRA
DONALD
PHILIP W.
STEVEN
NORTHERN KE

JOHN WILEY
New York   Ch

ISBN 0-471-53969-4

Printed in the United States of America

10 9 8 7 6 5 4 3 2 1

Introduction

To the student

In recent years several computer systems have appeared which are capable of performing the symbolic computations traditionally taught in algebra and calculus courses. This manual is designed to introduce you to one of the most powerful of these, the graphics/symbolic computation program *Mathematica*™. Its use can enhance your understanding of both the fundamental concepts and the computational techniques in calculus.

The manual is written assuming that you have access to an Apple Macintosh™ computer running the *Mathematica* program, Version 2.0. The main *Mathematica* program runs on a wide variety of computer systems, but special user interfaces or "front-ends" make the versions for the Macintosh and Next™ computers especially convenient for student users. (Similar "notebook interfaces" are now available for 386 and 486 compatibles with Windows 3.0™.) We have tried to minimize references to the Macintosh front-end, but in cases where its features require special instructions these have been provided. Your teacher should provide you with directions for using *Mathematica* on the computers at your school.

This manual is designed to be used as a supplement to a traditional calculus textbook. We have used traditional examples and exercises to illustrate how *Mathematica*, through graphics and animation as well as symbolic and numerical computation, can provide new insights into the basic ideas of calculus. Many of the exercises have been chosen from the new edition of Anton's calculus text [1], and in these cases the numbering of the exercises in that textbook is indicated. The manual is not intended as a general guide to *Mathematica*, but rather as a guidebook for exploring *calculus*. We introduce only a small subset of the more than eight hundred *Mathematica* commands, and we make no attempt to teach *Mathematica* programming. An Index lists all the *Mathematica* commands and Macintosh features we use, together with the page numbers of their first few occurrences. Additional information on *Mathematica* is available in the user's manual [2], written by the leading developer of the *Mathematica* program.

The early chapters of the workbook introduce several of the most basic *Mathematica* commands in a simple "interactive dialog" format. We present a sequence of commands for you to type and execute, together with a brief commentary. It is important that you become familiar with the syntax of *Mathematica* commands and the effects of common typing errors. To encourage you to actually execute the commands, the output from the computer is not shown except in those cases when its interpretation requires detailed discussion. To fully comprehend the commentary, therefore, you will need to execute the *Mathematica* commands and examine the resulting output on your computer screen.

Following the interactive dialog, examples similar to typical textbook exercises are worked out, again with explanatory comments. These worked examples are then followed by a small number of exercises. As might be expected, the sophistication of the reader regarding both calculus and *Mathematica* is assumed to gradually increase as the course progresses. In the later chapters the student who has learned to use *Mathematica* as an aid will be prepared to tackle exercises more difficult than those in traditional textbooks.

[1] Calculus, Fourth Edition, by Howard Anton, John Wiley & Sons, Inc., 1992.

[2] *Mathematica*: a system for doing mathematics by computer, by Stephen Wolfram, Second Edition, Addison-Wesley, 1991.

Contents

Chapter 13. Partial Derivatives   110

Chapter 14. Multiple Integrals   132

Chapter 15 Integrals over Curves and Surfaces   157

Index

Chapter 1 Introduction to *Mathematica*

Basic *Mathematica* Commands: `N`, `%`, `Simplify`, `Expand`, `Factor`, `Solve`, `Numerator`, `Denominator`, `Plot`, `FindRoot`, `Show`

The discussion below is designed to introduce you to some basic *Mathematica* commands. You should work through each of these examples using *Mathematica*. Type the commands (shown in boldface type) <u>exactly</u> as shown. Use capital letters when shown. All built-in *Mathematica* commands must begin with a capital letter, and when two or more words are run together to form a command name, like `PlotStyle`, the initial letters of each word are capitalized. After typing each command press the *enter* key (<u>not</u> the *return* key) to execute the given command. Feel free to experiment, making up related problems of your own to see how the *Mathematica* commands work.

§1 Arithmetic

`3 + 4`

`234/561 + 23/97`
Mathematica can do arithmetic with fractions.

`5 * 4`

`5  4`

`5(3 + 1)`
*Multiplication can be indicated in Mathematica by an asterisk *, as in many computer languages. But it also is indicated by a space or just by juxtaposing the two factors, if no ambiguity results.*

`4 ^ 3`
The caret '^' indicates exponentiation.

`Sqrt[4]`
Mathematica commands use parentheses ( ), brackets [ ] and braces { } in distinct ways. In typing the following commands be careful to distinguish between these different grouping symbols.

`Sqrt[5]`

`5^(1/2)`
Mathematica leaves square roots unevaluated if they are irrational. To obtain a decimal approximation, enter

`N[%]`
The symbol % is used to refer to the previous output, while **N** *asks for a numerical approximation.*

`N[Sqrt[5],40]`
This produces an approximation of the square root of 5 to 40 digits.

`Sqrt[5.0]`
When a decimal value is used as input, Mathematica assumes that an approximate decimal answer is desired.

§2 Basic Algebra

`3x + 5x`
Since we have not assigned a value to x at this time Mathematica leaves the simplified expression 8x unevaluated.

`x = 5`
`3x + 5x`
Now Mathematica can evaluate the expression using the given value for x.

`Clear[x]`
This command clears the computer's memory of any definitions previously associated with the symbol x, making it available again as a name for a variable.

`3(x+5)  -  4(2x-4)^2`
The algebraic expression is simply echoed back unsimplified.

`Simplify[3(x+5)  -  4(2x-4)^2]`

`Simplify[x/(x^2 - 4x) + 2x/(3x^2 - 48)]`

The `Simplify` *command will prove very handy.*

Worked Examples. These examples introduce more *Mathematica* commands while also showing how they can be used to solve typical mathematics problems. Work through each of these examples using *Mathematica*.

Example 1 Multiply out the expression $(1-x)^2(x-3)(x+4)$.
Solution:

```
bob = Expand[ (1-x)^2  (x - 3) (x + 4) ]
```
We give a name to the expanded expression so that we don't have to retype it if we need to refer to it later. To avoid possible conflicts with names of Mathematica's built-in commands, we use a lower case letter as the initial letter of any names we assign to expressions. Note that the extra spaces we used to make the algebraic expression easy to read were ignored by Mathematica. When you type complicated Mathematica expressions, use plenty of spaces to make their structure clear.

Example 2 Factor the expression you obtained in Example 1.
Solution:

```
Factor[bob]
```
A piece of cake!

Example 3 Find the roots of the same expression.
Solution:

```
Solve[bob == 0, x]
```
Note the double equal sign, which indicates an equation to be solved. Since an equation may involve several variables, we explicitly tell Mathematica that we wish to solve for x.

Example 4 Solve the equation $x^{12} - 7x^4 + 7x + 9 = 0$.
Solution:

```
ralph = x^12 - 7x^4 + 7x + 9

Solve[ralph == 0, x]
```

3

Since polynomials of degree 5 or more usually can't be solved algebraically, we must be content with decimal approximations to the solutions:

```
N[%]
```
Notice that the twelfth degree polynomial ralph has twelve roots. Mathematica finds complex roots as well as real roots.

Example 5   Factor the numerator of the expression $\dfrac{9x^2 - 18x + 8}{x^3 + 8x^2 - 9x - 72}$, and find the roots of the denominator.

<u>Solution:</u>

```
sue = (9x^2 - 18x + 8)/(x^3 + 8x^2 - 9x - 72)

Factor[Numerator[sue]]

Solve[Denominator[sue] == 0, x]
```

Example 6   Simplify the expression $\dfrac{x}{\sqrt{x^2 - 7}} + \sqrt{x^2 - 7}$.

<u>Solution:</u>

```
Simplify[ x/Sqrt[x^2 - 7] + Sqrt[x^2 - 7] ]
```
Easy as that!

Exercises

1)   Multiply the polynomial $x^3 + 4x + 9$ by the polynomial $4x^2 - 4x - 3$.

2)   Find all points where $x^3 - 11x^2 - 5x + 55 > 0$, by factoring the polynomial and then using a "sign chart" to record the intervals on which the individual factors have constant sign. (See [1], Example 1, page 217 for an example.)

3)   Solve $\dfrac{x - 3}{\sqrt{x - 8}} - 9x = 0$. Find decimal approximations of the real solutions.

4)   Solve $x^8 - 2 = 0$. How many solutions are real? Find decimal
      approximations of the real solutions.

§3 Plotting Commands

This section introduces the commands used to plot graphs of functions of one
variable.

Example 1 Plot the graph of $y = x^2 - 9$, for x between –5 and 5.
<u>Solution:</u>


```
Plot[ x^2 - 9, {x,-5,5}]
```
*The scale for the x-axis is different than that for the y-axis. If we want to have equal scales we
can set the the* AspectRatio *to* Automatic.


```
Plot[ x^2 - 9, {x,-5,5}, AspectRatio -> Automatic]
```
As we shall see in the next chapter, AspectRatio *is one of many options available for use in
the* Plot *command.   (Note that the* -> *arrow is created by typing a hyphen – followed by the
"greater than" symbol* >*, with no space between the two.)*

Example 2 Plot the graph of $y = \dfrac{x^2 - 1}{x^3}$, showing all interesting features of the
graph as clearly as possible.
<u>Solution:</u>


```
Plot[(x^2 - 1)/x^3, {x,-5,5}, AspectRatio -> Automatic]
```
Because we have kept the units on each axis the same by using the
AspectRatio->Automatic *option, the graph we obtain is not a good one. Let's plot the
function again, this time not specifying the* AspectRatio:


```
Plot[(x^2 - 1)/x^3, {x,-5,5}]
```
Here and later you can save tedious typing, by using the **Copy** *and* **Paste** *commands (in the*
Edit *menu) to make a copy of the previous* **Plot** *command , and then make the required
changes in the copies.*

From the second plot we see the general shape of the graph, but the scale on the y-axis is so coarse that much of the graph appears to coincide with the x-axis. To correct this we will use the **PlotRange** option, which allows us to specify the range of values shown on the y-axis. This is one of the most useful plotting options.

```
Plot[(x^2  -  1)/x^3,{x,-5,5},PlotRange->{-5,5}]
```
This gives a much better picture of the graph. Let's reduce the PlotRange even further.
```
Plot[(x^2  -  1)/x^3,{x,-5,5},PlotRange->{-1,1}]
```
This last plot clearly shows the interesting features of the graph.

Example 3 Computer disks cost $1.23 each if bought in small quantities. They sell for $0.99 each if bought in quantities of 100 or more. Express the cost of buying n disks as a function of n, for $0 \leq n \leq 500$. Graph this cost function.
Solution:

For under 100 disks, the cost of buying n disks will be 1.23n dollars. For 100 or more it will be 0.99n dollars. Thus the cost function is given by

$$c(x) = \begin{cases} 1.23x & \text{if} \quad 0 \leq x \leq 99 \\ 0.99x & \text{if} \quad x \geq 100 \end{cases}$$

We can plot c(x) by combining the plots of its two component functions, using Mathematica's Show command:
```
plot1  =  Plot[  1.23x,{x,0,99}]
```

```
plot2  =  Plot[  0.99x,   {x,100,500}]
```

```
Show[plot1,plot2]
```
(Note: Even though the function c(x) is meant to be evaluated only for integer values of x in this applied problem, we have drawn its graph as if x were allowed to be any real number.)

Example 4 Use graphs to show that $\cos x = x$ has only one solution. Find this solution accurate to three digits.
Solution:

First we use graphs to visualize the problem.

Plot the graphs of $y = \cos x$ and $y = x$ on the same coordinate plane. The x-coordinate of any points of intersection of the two graphs is a solution of the equation $\cos x = x$. (Note that to

6

plot two or more functions together we must list them between braces { } as shown below.)
```
Plot[{Cos[x],x},{x,-3,3}]
```

We see a single intersection, near the point (0.7,0.7). To estimate the coordinates of this point more accurately, we can use a special feature of the Macintosh version of Mathematica. Click the mouse on the plot to obtain a box around it. Then when you hold down the "apple" key (marked with the symbols ⌘ and ⌘) the cursor turns into crosshairs. Place the crosshairs on the intersection point and the coordinates of the crosshairs will appear in the lower left corner of your Mathematica window.

To get a better approximation we re-plot over a smaller interval of x-values.
```
Plot[{Cos[x],x},{x,0.70,0.75}]
```

Again using the crosshairs, we estimate the intersection point to be (0.739, 0.739). Check this:
```
Cos[0.739]
```
This is not exactly 0.739, but the two numbers agree to 3 decimal places. So we conclude that the equation $\cos x = x$ *has only one solution, approximately* $x = 0.739$.

Mathematica has a special command, **FindRoot**, *for improving upon an initial numerical approximation to a solution of an equation.*

```
FindRoot[ Cos[x] == x, {x,0.7}]
```
The x in the list brackets is the variable, and the number 0.7 is the point at which to begin looking for a nearby solution (root).

When the graphs of two or more functions are drawn in a single plot, it is useful to distinguish between graphs by displaying them in different colors. This can be accomplished by using the **PlotStyle** option.

Example 5 Plot the function $y = x^5 - 7x + 4$ in green.
Solution:

```
Plot[ x^5 -7x +4, {x,0,5}, PlotStyle->RGBColor[0,1,0] ]
```

The first number after **RGBColor** *is the amount of red, second is green and the third is blue (hence RGB.) Each number must be between zero and one. Before proceeding, experiment by replacing the pure green* **RGBColor[0,1,0]** *specification in this command by various RGB combinations of your own. For example try* **RGBColor[1,1,0]** *and* **RGBColor[0,0,0]**.

Example 6 Plot the two functions $y = \sin x$ and $y = \sin 2\pi x$ for $0 \leq x \leq 2\pi$, using a different color for each graph.

Solution:

```
Plot[{Sin[x],  Sin[2Pi  x]},{x,0,2Pi},
        PlotStyle->{ RGBColor[1,0,0],  RGBColor[0,1,0]}]
```

The graph of the first function $\sin x$ *appears in red and the second function* $\sin 2\pi x$ *is shown in green. (Carefully note the capital letters in* **Sin**, **Pi** *and* **RGBColor**, *and the way brackets [ ] and braces { } are used in this command. Also note that a carriage return was used to break the command into two lines at a convenient place, and tabs or spaces were inserted at appropriate points to make the command easy to read. Try to develop a habit of making your Mathematica commands easy to follow for a reader.)*

It is difficult to remember the **RGBColor** recipe for common colors, so the Macintosh version of *Mathematica* provides a **Color Selector** which you can use to get the **RGBColor** specification for any color you like.

Example 7 Use the **Color Selector** to plot the graph of $y = x^2$, for $-5 \leq x \leq 5$, in purple.

Solution:

First type
```
Plot[x^2,{x,-5,5},PlotStyle->  ]
```
leaving the cursor between the arrow -> *and the bracket], where you want the* **RGBColor** *specification for purple to appear. Then use the color wheel to paste this* **RGBColor** *recipe into your command as follows:*

> a) *Pull down the* **Action** *menu until* **Prepare Input** *is highlighted.*
> b) *Holding down the mouse button, carefully slide the cursor to the side into the pop up menu and choose* **Color Selector...**. *Release the mouse button.*
> c) *The color wheel will appear. Click on the shade of purple you want.*
> d) *Click on* **OK**. *The* **RGBColor** *recipe for the color you selected will appear in your Mathematica command like this:*

```
Plot[x^2,{x,-5,5},PlotStyle->RGBColor[0.659,  0.045,  0.974]  ]
```

Example 8 Solve $\dfrac{x^2+9}{3x+8} > 4x^2+7$.

Solution:

We begin by making a plot of the functions on the two sides of the inequality. But what interval of x-values is appropriate for the plot? Since x^2+9 and $4x^2+7$ are always positive, any solution of our inequality must satisfy $3x+8>0$. That is, $x > -\frac{8}{3}$. Let's examine a plot over the interval $-\frac{8}{3} \le x \le 10$:

```
Plot[ {(x^2 + 9 ) / (3x + 8), 4x^2 + 7}, {x,-8/3,10},
        PlotStyle->{RGBColor[1,0,0],RGBColor[0,0,1]}]
```

The inequality is satisfied wherever the red curve lies above the blue one. From the graph we see that this happens only on a small interval near –2. To better identify the endpoints of this interval we re-plot the two functions over the interval $[-\frac{8}{3}, -2]$. We will also limit the plot range to the interval $[0, 50]$, which the plot above shows to be sufficient.

```
Plot[ {(x^2 + 9 ) / (3x + 8), 4x^2 + 7}, {x,-8/3,-2},
      PlotStyle->{RGBColor[1,0,0],RGBColor[0,0,1]},
      PlotRange->{0,50}  ]
```

From the plots we see that the solution is approximately $(-\frac{8}{3}, -2.5)$. To get a more accurate value for the right endpoint we solve the equation $\dfrac{x^2+9}{3x+8} = 4x^2+7$, using the **FindRoot** command.

```
FindRoot[(x^2 + 9 ) / (3x + 8) == 4x^2 + 7,{x,-2.5}]
```
So the solution to the inequality is $-8/3 < x < -2.5082$.

Exercises

After working each exercise explain your reasoning in a short paragraph. Use complete sentences. Explain how you went about finding your answer as well as your interpretation of each output.

1) Plot the graph of $f(x) = \dfrac{(x-2)^3}{x^2}$. Find a plot which shows the interesting features of the graph of this function.

2) Solve $x^3 - 5x + 2 > 0$, by first using the graph of $y = x^3 - 5x + 2$ to estimate the endpoints of the solution intervals, and then solving the equation $x^3 - 5x + 2 = 0$ (using the **Solve** command) to find the endpoints exactly.

3) Solve $2^x > x^4$. Find the endpoints of the solution intervals accurate to two decimal places using graphical methods; then use the **FindRoot** command to find them more accurately.

4) Plot the graphs of $y = x$, $y = -x$ and $y = x \sin x$ together, over the interval $-2\pi \le x \le 2\pi$. Use the **Color Selector** to color the first two graphs brown and the third turquoise. (Recall that π is named **Pi** in *Mathematica*.)

5) Use graphical methods to approximate the solution of $|2x - 1| > |x^2 - 7|$. (Note: *Mathematica*'s absolute value function is named **Abs[ ]**.) Then use appropriate **Solve** commands to find the endpoints of the solution intervals exactly. (Find equations for the curves, near each crossing point, which do not involve the absolute value function.)

6) Plot the function $y = x^5 - 8x^4 + 11x^3 + 56x^2 - 180x + 144$. You will need at least two plots in order to show all of the interesting features of the graph of this function.

§4 Parentheses and Brackets

Mathematica distinguishes between parentheses and different types of brackets. Understanding what each is used for will help you avoid many mistakes in formulating your own *Mathematica* commands.

( ) Parentheses are used for grouping.

Examples: `(3x + 4y)(4x + 9y),    3^(1/3),    2x/(x + 1)`

[ ] - Square brackets are used to enclose the expressions acted upon in *Mathematica* commands and built-in functions, as well as functions defined by the user.

Examples: `N[Pi],    Sin[x],    Factor[x^2 - 3x + 2],    f[x_]`

{ } - Braces (also called <u>list brackets</u>) are used to enclose lists of elements. Many *Mathematica* commands allow the user to replace a single element by a list of elements.

Examples:

i) `Plot[{f[x],g[x]},{x,1,10}]`   The first pair of braces contains a list of functions (rather than the usual single function), the second braces enclose a list indicating the variable and the interval of values over which the graphs of the functions are to be plotted.

ii) `Solve[{2x + 5y == 7, 4x - 3y == 12}, {x,y}]`
The first pair of braces contains a list of equations, and the second pair of braces contains the list of variables for which the equations are to be solved. The output of this command is a list of solutions. Since the given pair of equations has only a *single* solution $x=\dfrac{81}{26}$, $y=\dfrac{2}{13}$, the output is a list $\{\{x->\dfrac{81}{26},y->\dfrac{2}{13}\}\}$ whose only entry is the list of replacements $\{x->\dfrac{81}{26},y->\dfrac{2}{13}\}$.

Chapter 2 - Functions and Limits

New *Mathematica* Commands:	`f[x_]`
	Table
	Limit

§1 Defining Functions

Mathematica allows you to define functions and use them in much the same way they are used in mathematics. In this section we will learn how to define functions and use them in various ways.

Example 1 Define the function $f(x) = 3x + 6$ and find the value of the function at 2 and at -3.5. Plot the graph of f over the interval $-1 \leq x \leq 4$.
<u>Solution:</u>

`f[x_] = 3x + 6`
Note the underline "_" next to the x. This tells Mathematica that the "x" is a variable and can hence be replaced by any other expression. A common error is to leave out the underline, in which case Mathematica won't treat f as a function.

`f[2]`
Notice that the computer "remembers" the definition of the function.

`f(-3.5)`
Remember that functions require square brackets, not parentheses!
`f[-3.5]`

`f[a+b]`

`Plot[f[x],{x, -1, 4}]`
Note that we use just x here, not x_. <u>The underline after the x is used only in the left-hand side of the original definition of the function f</u>.

12

Example 2 Define the function $g(t) = t^2 - 7$. Find $g(x)$, $g(u)$, $g(x+h)$, and $g(anything)$.

Solution:

```
g[t_] = t^2 - 7
```

```
g[x]
```

```
g[u]
```

```
g[x+h]
```

```
g[anything]
```

Note how the variable used does not matter. Because of this we often call the variable used to define a function a "dummy" variable. The function g is best described as the function which squares its input and then subtracts seven.

Example 3 Define a function which takes two inputs and adds them together. Use this function to add 45 and 67. Then use it to add apples and oranges.

Solution:

```
add[x_,y_] = x + y
```

```
add[45,67]
```

```
add[apples,oranges]
```

Exercises

1. Define a function which squares its input and then divides the result by 56. Find the value of this function at $x = 14$.

2. a) Define g to be the function which gives as output the square of its input, plus the input, plus 7.
   b) Find g[x+h] and then use the computer to simplify the expression (g[x+h] - g[x]) / h.

3. a) Define a function f whose output is twice the sine of the input.

13

b) Define a function *g* which takes the square root of one less than its input.

c) Define a function *h* which is *g* plus twice *f*, where *f* and *g* are the functions defined in parts a) and b) above. Find the domain of *h* and plot the graph of *h*. Be sure that your plot shows all of the interesting features of h.

§2 Limits

We will describe several ways to analyze limits using *Mathematica*.

Example 1 Investigate $\lim_{x \to 0} \dfrac{1}{x}$ graphically.

Solution:

```
Clear[f]
```
This will clear any previous definitions associated with the name f from the computer's memory.

Now define $f(x) = \dfrac{1}{x}$ *as a Mathematica function:*

```
f[x_]  =  1/x
```
Plot the function over an interval centered at x = 0:

```
Plot[f[x],{x,-1,1}]
```

We see that as x approaches zero from the left, f(x) gets very large negative (that is, large in absolute value, but negative). We may express this concisely by saying the limit of f as x approaches zero from the left is negative infinity, or by writing $\lim_{x \to 0^-} \dfrac{1}{x} = -\infty$. *As x approaches zero from the right the value of f(x) grows very large positive:* $\lim_{x \to 0^+} \dfrac{1}{x} = \infty$. *Since the two one-sided limits are not equal the limit* $\lim_{x \to 0} \dfrac{1}{x}$ *does not exist. (Thus if we are told that x is very near zero but are not told which side of zero it is on, we <u>cannot</u> predict the value of f(x) — all we can say is that f(x) will have a very large absolute value.)*

Example 2 Investigate the limit: $\lim_{x \to 4} \dfrac{x^2 + 2x - 24}{x^3 + 12x^2 - x - 252}$.

Solution:

```
Clear[g]
```

```
g[x_]  =  (x^2 + 2x - 24)/(x^3 + 12x^2 - x - 252)
```

14

```
Plot[g[x],{x,2,6}].
```
From the graph the limit should be about 0.07.

```
g[4]
```
Why do we get the response Indeterminate *when we try to evaluate g(4)? The graph looks nice enough. Actually there is a gap in the graph of g where x=4, since the numerator and denominator are both zero when x = 4, but the plot produced by Mathematica fails to show this gap. Keep in mind that a Mathematica plot is only an approximation to the exact graph.*

We can approximate the limit numerically by examining the values of g(x) for a sequence of values of x approaching 4 from the left, and for a sequence approaching 4 from the right.

```
leftpoints = Table[N[4 - 10^(-k),30], {k,1,10} ]
```
This Table *command creates the list* $\{4-10^{-1}, \ 4-10^{-2}, \cdots, \ 4-10^{-10}\}$, *each accurate to 30 digits. Mathematica allows functions to act on lists of numbers, applying the function to each entry in the list:*

```
leftvalues = g[leftpoints]
TableForm[leftvalues]
```
Scanning this table of values of g(x) as x approaches 4 from the left, it seems clear that the values are approaching a number whose decimal expansion begins 0.0699300.... *Thus it appears that* $\lim_{x \to 4^-} g(x)$ *exists and is this number. Now we investigate the limit from the right:*

```
rightpoints = Table[N[4 + 10^(-k),30], {k,1,10} ]
rightvalues = g[rightpoints]
TableForm[rightvalues]
```
The left-hand and right-hand limits of g(x) at 4 appear to be the same, so it appears from our numerical investigation that $\lim_{x \to 4} g(x) = 0.0699300....$

    In this example we can find the limit exactly, by making use of Mathematica's symbolic algebra commands. Both the numerator and denominator of g(x) are zero when x = 4. Whenever the numerator and denominator of a rational function have a common root, the function can be algebraically simplified.

```
Simplify[g[x]]
```

```
    6 + x
--------------

              2
63 + 16 x + x
```

(This is the Mathematica output generated by the **Simplify** *command.)*
As x → 4 this clearly approaches

```
(6+4)/  (63  +  16*4  +  4^2)
```

So the limit is exactly $\dfrac{10}{143}$.

```
N[%,10]
```

Example 3 Investigate $\displaystyle\lim_{x\to 0}\sin\!\left(\frac{1}{x}\right)+1$.

<u>Solution:</u>

```
Clear[f]
```
This removes any old definitions associated with the name f.

```
f[x_]  =  Sin[1/x]  +  1
```

```
Plot[f[x],{x,-2,2}]
```
Does the limit exist? Suppose someone claimed that the limit is one. Does the value of f(x) get near one as x gets near zero? Yes, but does it stay there? No, so the limit doesn't equal one. As x approaches zero the values of f(x) pass through every value from 0 to 2, but do not remain close to any one of these numbers. Hence the limit does not exist.

The same conclusion can also be reached by investigating the limit numerically, as in Example 2:
```
rightpoints  =  Table[N[0  +  10^(-k),30],   {k,1,10}  ]
rightvalues  =  TableForm[f[rightpoints]]
```
The values of f(x) do not appear to approach a limit as x approaches 0 from the right. Therefore $\displaystyle\lim_{x\to 0^{+}} f(x)$ *does not exist, and therefore the (2-sided) limit of f at 0 cannot exist.*

Example 4 Investigate the limit: $\displaystyle\lim_{x\to 0} x\sin\!\left(\frac{1}{x}\right)$.

<u>Solution:</u>

```
Clear[g]
```

```
g[x_]  =   x Sin[1/x]
```

```
Plot[g[x],{x,-1,1}]
```
From the plot, the value of g(x) appears to approach zero as x approaches zero. Since the factor

Sin[1/x] oscillates between –1 and 1, the graph of g lies between the lines y = x and y = –x.

```
Plot[{x,  -x,  g[x]},   {x,-1,1},   PlotStyle ->
    {RGBColor[1,0,0],RGBColor[1,0,0],RGBColor[0,0,0]},
    AspectRatio -> Automatic ]
```

Thus if x is near zero, so is the value of g(x). We conclude that $\lim\limits_{x \to 0} x \sin\left(\dfrac{1}{x}\right) = 0.$

This conclusion can be quickly verified by examining tables of numerical values as described in Examples 2 and 3 above.

Mathematica has a command for finding limits, but it occasionally gives incorrect results or is unable to reach a conclusion. Thus limits computed by using the **Limit** command should be checked by alternative methods.

Example 5 Use *Mathematica*'s **Limit** command to investigate the following limits. Check each answer.

a) $\lim\limits_{x \to 2} \dfrac{x^2 + 2x - 8}{x^2 - 8x + 12}$

b) $\lim\limits_{x \to +\infty} \dfrac{x^2 + 2x + 9}{x^2 + 8x + 12}$

c) $\lim\limits_{x \to -2} \dfrac{1}{x^2 + 9x + 14}$

Solution:

a)
```
Clear[f,g,h]
f[x_]  =  (x^2 + 2x - 8)/(x^2 - 8x + 12)
Limit[f[x],  x -> 2]
```
Mathematica says the limit is –3/2. (Recall that the arrow -> is a hyphen – followed by a "greater than" symbol >.)

We can check this using a table of values:
```
leftpoints = Table[ N[2 - 10^(-n),20],{n,1,10}]
leftvalues = TableForm[f[leftpoints]]
```

17

We see that from the left the values approach –1.5. We leave it to the reader to check that values from the right also approach –3/2.

b) `g[x_] = (x^2 + 2x + 9)/(x^2 + 8x + 12)`

`Limit[g[x],x -> Infinity]`
Mathematica says the limit is 1. Note that Mathematica's name for ∞ is the word Infinity.

We can again check the limit with a table of values:
`points = Table[ N[10^n, 20],{n,5,15}]`
`values = Tableform[g[points]]`
From the table it seems that Mathematica was correct, the limit does appear to be 1.

c) `h[x_] = 1/(x^2 + 9x + 14)`
`Limit[h[x], x -> -2]`
Mathematica says the limit is Infinity.

This time we will check the limit with a plot:
`Plot[h[x], {x,-3,-1}]`

From the plot we see that the limit as calculated by Mathematica is incorrect. However Mathematica does find the one-sided limits correctly:

`Limit[h[x], x -> 2, Direction -> -1]`
The option `Direction -> -1` *tells Mathematica to approach the point –2 from the <u>right</u>, i.e., going in the direction one moves on the number line to reach –1, starting from 0.*

To find the limit from the left we specify `Direction -> 1`
`Limit[h[x],x -> 2,Direction -> 1]`

We conclude that the limit doesn't exist, but that the limits from the left and from the right exist and are –∞ and ∞, respectively.

Mathematica note: The choice of values for the `Direction` *option is confusing, since it seems to conflict with the standard mathematical terminology for one-sided limits in calculus. But if you remember that Mathematica is designed to deal with complex numbers as well as real numbers, the specification of a direction by <u>telling which way to go from 0</u> makes sense.)*

Exercises

Investigate each of the limits in problems 1-5
   a) by graphing,
   b) numerically, as in Example 2 above,
   c) through the use of *Mathematica's* `Limit` command.
For each exercise explain your conclusions, in complete sentences.
(For example in describing the conclusions reached in Example 5 c) above, you might write: "The limit command gave an answer of infinity. However when we used the commands for one-sided limits, we found that the limit from the left is negative infinity the limit from the right is infinity . Thus *Mathematica's* result was incorrect: the limit is not ∞, instead the limit does not exist.")

1) (§2.5, #10) $\lim\limits_{x \to 3} \dfrac{x^2 - 2x}{x + 1}$

2) $\lim\limits_{x \to 7} \dfrac{x^4 - 14x^3 + 48x^2 + 14x - 49}{x^5 - 7x^4 - 5x^2 + 39x - 28}$

3) $\lim\limits_{x \to -4} \dfrac{x^6 + 4x^5 - 2x - 8}{x^3 + 4x^2 - 9x - 36}$

4) $\lim\limits_{x \to 2} \dfrac{\sin\left(\dfrac{x - 2}{3}\right)(x^3 - 6x^2 - 24x + 64)}{x^6 - 6x^5 + 9x^4 + 17x^3 - 78x^2 + 108x - 56}$

(Remember that the `Limit` command gives some incorrect answers.)

5) (§2.5, #27) $\lim\limits_{y \to -\infty} \dfrac{2 - y}{\sqrt{7 + 6y^2}}$

6) $\lim\limits_{x \to 0} \dfrac{\cos(x)}{x}$ (Don't put too much faith in the `Limit` command!)

19

§3 Continuity

Example 1 Which of the following functions is continuous at $x = 0$? (Recall that f is continuous at 0 if $f(0)$ is defined, $\lim_{x \to 0} f(x)$ exists and $\lim_{x \to 0} f(x) = f(0)$.)

a) $f(x) = \dfrac{1}{x}$

b) $h(x) = \begin{cases} 0 & \text{if } x = 0 \\ \dfrac{x}{|x|} & \text{if } x \neq 0 \end{cases}$

c) $g(x) = \begin{cases} 0 & \text{if } x = 0 \\ x \sin\left(\frac{1}{x}\right) & \text{if } x \neq 0 \end{cases}$

Solution:

a) Since $f(0)$ is undefined, f is not continuous at $x = 0$. Also, by examining a plot of the graph over an interval containing $x = 0$, we see that f has no limit at zero and so can't be continuous there:

```
Plot[1/x,{x,-1,1}]
```

b) We define $h(x)$ as a Mathematica function, and plot its graph near $x = 0$:
```
h[x_]  =  x/Abs[x]
```

This defines our function except at the single point $x = 0$.
```
Plot[h[x],{x,-1,1}]
```

We see that the value $h(x)$ jumps from -1 to the left of 0 to $+1$ to the right of 0. Thus the left-hand and right-hand limits at 0 are not equal, so $\lim_{x \to 0} h(x)$ does not exist. Therefore h is not continuous at 0.

c) We examined the behavior of $g(x) = x \sin\left(\dfrac{1}{x}\right)$ near 0 in Example 4 of §3 above. Our conclusion was that $\lim_{x \to 0} x \sin\left(\dfrac{1}{x}\right) = 0$. Thus the limit of g at 0 equals the value $g(0)$, so we conclude that g is continuous at the point $x = 0$.

Example 2 Choose values for *a* and for *b* so that the function defined below is continuous.

$$f(x) = \begin{cases} x^3 + 9 & \text{if} \quad x \le 4 \\ ax + b & \text{if} \quad 4 < x \le 9 \\ 2\sin(3x) & \text{if} \quad x > 9 \end{cases}$$

Solution:

Since each of the function's "component parts" are continuous we need only be sure the function is continuous at the points x = 4 and x = 9 where the rule defining the function changes. Because the limits of the component functions $x^3 + 9$, $ax + b$ and $2\sin 3x$ at 4 and 9 equal their values there, we just need to make sure that the value of $x^3 + 9$ equals the value of $ax + b$ at x = 4, and that $ax + b$ equals $2\sin 3x$ at x = 9. So the unknown numbers a and b must satisfy the equations

$4^3 + 9 = 4\,a + b$

$9a + b = 2\,\sin(\,3*9)$

First we solve these equations:

```
Solve[{4^3 + 9 == 4 a + b,9 a + b == 2 Sin[27]}]
```

```
N[%]
```

Thus a = −14.2174 and b = 129.87 are approximate values of a and of b. To check our work we plot the graph of the resulting function f (using the approximate values for a and b) :

```
plot1 = Plot[x^3 + 9,{x,0,4}]
plot2 = Plot[-14.2174 x + 129.87,{x,4,9}]
plot3 = Plot[2 Sin[3x],{x,9,12}]
```

```
Show[{plot1,plot2,plot3}]
```
The plot shows no jump discontinuities.

Exercises

1) (§2.7, #17b) Find a value for the constant k that will make the function f continuous:

$$f(x) = \begin{cases} kx^2, & x \leq 2 \\ 2x + k, & x > 2 \end{cases}.$$

Plot the resulting function to check its continuity.

2) At which values of x are each of the following not continuous? Carefully describe the nature of each discontinuity.

a) (§2.7, #10) $f(x) = \dfrac{3x+1}{x^2 + 7x - 2}$

b) (§2.7, #14) $f(x) = \dfrac{x+3}{\left| x^2 + 3x \right|}$

3) Is $f(x) = \begin{cases} \dfrac{\sin(x^2)}{x^2} & \text{if } x \neq 0 \\ 1 & \text{if } x = 0 \end{cases}$ continuous at $x = 0$? Explain.

22

Chapter 3 - The Derivative

```
                          f', f"

                          D

New Mathematica Commands:  ContourPlot

                          FindRoot

                          ReplaceAll
```

§1 The Definition of the Derivative

These examples and exercises utilize *Mathematica* to review the definition of the derivative from several points of view.

Example 1 Use graphical methods to estimate the slope of the graph of the function $f(x) = 3 + 2sin(\pi x)$ at the point (1,3).

Solution:

Define f and plot its graph near x = 1:
```
f[x_] = 3 + 2 Sin[Pi x]
```

```
Plot[f[x],{x,0,2}]
```
Because we can't find the slope of a curve (without computing the derivative) we will plot the graph over smaller and smaller intervals centered at x = 1, until the graph looks like a straight line:

```
Plot[f[x],{x,0.5,1.5}]
```

```
Plot[f[x],{x,0.9,1.1}]
```
This last plot looks rather straight. We choose (1, 3) and a nearby point on the graph (found using the cursor crosshairs as described in Chapter One), and compute the slope of the secant line through these two points using the usual rise over run definition. For example when x = 1.05 the y-coordinate on the latest graph appears to be 2.69. So we take (1.05, 2.69) as our second point:

```
slope = (3 - 2.69)/(1 - 1.05)
```

23

Compare this with the exact slope at (1, 3), given by the derivative:

```
f'[1]
```

```
N[%]
```

Our graphical approximation was fairly accurate Observing that the graph of a function over sufficiently short intervals appears to be a straight line is one way of seeing that the function is differentiable. Graphical methods are very useful for interpreting many ideas in calculus, but the results are often of limited accuracy so they must be supplemented with computations, as in the following example:

Example 2 For the function $f(x) = x^3 - 3x + 4$,
i)   Plot the graph of f and its tangent line at the point $(4, f(4))$.
ii) Find the three secant lines through $(4, f(4))$ and $(4 + h, f(4 + h))$ for $h = 0.5$, $h = 0.2$, and $h = 0.1$. Combine the graph of f, its tangent line and the three secant lines in a single plot.
iii) Verify that as $h \to 0$ the slopes of these secant lines approach that of the tangent line.

<u>Solution:</u>

i) First find the equation of the tangent line:

```
f[x_] = x^3 - 3x + 4
```

```
f'[4]
```
This gives the slope of the tangent line. Now use the point-slope form of a line:
$y - f(4) = f'(4)(x - 4)$ *to define the tangent line function:*

```
tangent[x_] = f[4] + f'[4] (x - 4)
```

Plot the function and its tangent line over an interval containing 4.
```
Plot[{f[x],tangent[x]},{x,3,5}]
```

ii) Now find equations for the secant lines:

```
slope1 = (f[4.5] - f[4])/0.5
secant1[x_] = f[4] + slope1 (x - 4)
```

*(Use **Copy** and **Paste** to simplify the typing of the following commands:)*

```
slope2 = (f[4.2] - f[4])/0.2
secant2[x_] = f[4] + slope2 (x - 4)


slope3 = (f[4.1] - f[4])/0.1
secant3[x_] = f[4] + slope3 (x - 4)
```

Plot the graph of f together with the tangent and secant lines:
```
Plot[{f[x],tangent[x],secant1[x],secant2[x],secant3[x]},
     {x,3,5},
     PlotStyle -> {RGBColor[1,0,0],RGBColor[0,1,0],
           RGBColor[0.01,  0.97,  0.95],
           RGBColor[0.09,  0.66,  0.97],
           RGBColor[0.13,  0.09,  0.97]}  ]
```

iii) The graph of f is red and the tangent line green. The smaller the value of h the darker the blue of the corresponding secant line. (We used the ColorSelector to find the RGB values for these shades of blue, as described in Chapter 1.) The three secants approach the tangent, so their slopes must also approach the slope of the tangent.
```
slope1
```
```
slope2
```
```
slope3
```
```
f'[4]
```
As h decreases the slopes do appear to approach f'(4).

Example 3 For the function $f(x) = x - \sqrt{x+9}$,
i)   Plot the graph of the derivative f' over the interval $-8 \le x \le 2$.
ii) Plot the graph of the difference quotient $\dfrac{f(x+h) - f(x)}{h}$ as a function of x,
     over this interval, for each of the values $h = 1.0, 0.5,$ and 0.1.
iii) Combine the graphs of f' and the three difference quotients in a single plot, and interpret this plot.

Solution:

i) *Define the function f and compute its derivative:*
```
f[x_] = x - Sqrt[x+9]
```

25

```
f'[x]
```

We plot the graph of the derivative on [–8, 2] (Note that f(x) is defined for x ≥ –9, but f'(–9) does not exist.)

```
plot1  =  Plot[f'[x],   {x,-8,2}]
```

ii) *Repeat this procedure for the difference quotients:*

```
dq[x_,h_]  =  (f[x+h]  -  f[x])/h
```

```
plot2  =  Plot[{dq[x,1.0],dq[x,0.5],dq[x,0.1]},   {x,-8,2},
              PlotStyle->{RGBColor[1,0,0],RGBColor[0,1,0],
                          RGBColor[0,0,1]}  ]
```

iii) *Now combine the plots:*

```
Show[plot1,plot2]
```
We see that the graphs of the difference quotients converge rapidly to the graph of the derivative, as h → 0.

Example 4 Use the definition of the derivative to find the derivative of each function.

a) $f(x) = x^2 + 5$

b) $g(x) = \dfrac{x+4}{3-x}$

<u>Solution:</u>

a) *Enter the difference quotient* $\dfrac{f(x+h)-f(x)}{h}$ *for this function and take the limit as h → 0:*

```
Clear[dq]
dq = ( ((x+h)^2 + 5) - (x^2 + 5) ) / h
```

```
Simplify[dq]
```

As h → 0 this expression approaches the limit 2x. So $f'(x) = 2x$.

b) *To simplify our work we first define the function g:*

26

```
g[x_]  =   (x+4)/(3-x)
```

Then define the difference quotient in terms of g:
```
Clear[dq]
dq  =  (g[x+h]  -  g[x])/h
```

When working this problem with pencil and paper the most difficult step is simplifying the difference quotient. We give Mathematica that chore:
```
dq  =  Simplify[%]
```

It is now obvious that the limit as $h \to 0$ of the difference quotient is $g'(x) = \dfrac{7}{(x-3)^2}$.

We check:
```
g'[x]
```

```
Simplify[%]
```
This agrees with our answer above.

Example 5 Investigate the derivative of the function $f(x) = |x|$. Where is f differentiable? Find a formula for $f'(x)$.
Solution:

First we try the obvious:
```
f[x_]  =  Abs[x]
```

We ask Mathematica to compute the derivative:
```
f'[x]
```
No help there.

Plot the graph of Abs[x]:
```
Plot[Abs[x],{x,-1,1},   AspectRatio->Automatic]
```
(The option AspectRatio –> Automatic makes the scales equal on the two axes.)

*For x > 0 the slope is +1, and for x < 0 the slope is –1. Therefore f is differentiable for all $x \neq 0$
and $f'(x) = \begin{cases} -1 & \text{if } x < 0 \\ 1 & \text{if } x > 0 \end{cases}$.*

What about $f'(0)$?

We begin our investigation by plotting y = Abs[x] over increasingly smaller intervals, as in example 1:

```
Plot[Abs[h],{h,-1,1}]
```

```
Plot[Abs[h],{h,-.1,.1}]
```

```
Plot[Abs[h],{h,-.001,.001}]
```

We conclude from the plots that since the graph doesn't appear to be approaching a straight line that f is not differentiable at zero.

Let's look at the question of differentiabilty of Abs[x] at zero in another way:

Consider the difference quotient at x = 0: $\dfrac{f(0+h)-f(0)}{h} = \dfrac{|h|}{h}$.

Plot this function of h:

```
Plot[Abs[h]/h,{h,-1,1}]
```

We see that the limit as $h \rightarrow 0$ of the difference quotient at x = 0 doesn't exist (the limit from the left is –1 and the limit from the right is +1). Again we conclude that f'(0) doesn't exist.

A simple way to write our result is $\dfrac{d}{dx}(|x|) = \dfrac{|x|}{x}$, since the expression on the right is

$$\begin{cases} -1 & \text{if } x < 0 \\ 1 & \text{if } x > 0 \\ \text{undefined if } x = 0 \end{cases}$$

As a check we plot the difference quotient as a function of x, using small values for h, say h = 0.1 and h = 0.02. This should look very much like the plot of the graph of the derivative of |x|.

```
Plot[ (f[x+.1] - f[x])/.1, {x,-1,1} ]
```

```
Plot[ (f[x+.02] - f[x])/.02, {x,-1,1} ]
```

Compare these plots with the plot of the derivative $\dfrac{|x|}{x}$:

```
Plot[Abs[x]/x,{x,-1,1}]
```

We see that as $h \rightarrow 0$ (in our case h =.1, .02) the graph of the difference quotient as a function of x approaches the graph of the derivative. (Note: The curve drawn by Mathematica in plotting $\dfrac{|x|}{x}$ is slightly in error, since it includes a vertical segment joining (0, -1) and (0, 1), whereas $\dfrac{|x|}{x}$ is undefined at x = 0.)

Remark Actually, *Mathematica* <u>can</u> correctly differentiate the absolute value function **Abs[x]**, but only if we replace it by the equivalent algebraic expression **Sqrt[x^2]**. The command **D[Sqrt[x^2],x]** (which we'll see in §2 tells *Mathematica* to differentiate **Sqrt[x^2]** with respect to x) yields the output $\dfrac{x}{\text{Sqrt}[x^2]}$, i.e., $\dfrac{x}{|x|}$, which is equivalent to our earlier formula $\dfrac{|x|}{x}$. It is often useful to replace *Mathematica*'s **Abs[x]** function by **Sqrt[x^2]**.

Exercises

1. a) Use graphical methods to estimate the slope of $f(x) = 2^{x^2+1}$ at the point
        $P = (1,4)$.

    b) Find the exact slope at P.

    c) Find an equation for the line tangent to the graph of f at P. Plot the
        graph of the function and the tangent line at P together in a single plot.

2. a) Plot the graph of $g(x) = x \cos^2(x)$ along with its tangent line through the
        point (π, π).

    b) Create a plot showing the secant lines at (π, π) approaching the tangent
        line, for $h = 1.0$, $h = 0.5$, and $h = 0.1$.

    c) Find the slope of each of the secant lines in part (b) and show that as
        $h \to 0$ these slopes approach that of the tangent line.

3. a) Plot the graph of the derivative $g'(x)$, if $g(x) = x \cos^2(x)$.

    b) Plot the difference quotient $\dfrac{g(x+h) - g(x)}{h}$ as a function of x,

        for each of the values $h = 0.5$, $h = 0.2$, $h = 0.05$.

    c) Combine the plots in parts (a) and (b), and interpret the result.

4. Later in your course of study you will encounter the exponential function. In
    Mathematica this function is denoted **Exp[x]** .

    a) Plot the graph of the exponential function for x ranging from –5 to 5.

    b) Plot the difference quotient for the exponential function as a function of
        x, using a small value for h. From this plot what do you guess to be the
        derivative of the exponential function?

    c) Check your guess of part (b) by using *Mathematica* to find the derivative
        of the exponential function.

5. Investigate the differentiability at $x = 0$ of $f(x) = \begin{cases} x\sin(\frac{1}{x}) & \text{if } x \neq 0 \\ 0 & \text{if } x = 0 \end{cases}$.

6. Use the definition of the derivative to find the derivative of each of the following functions:

   a) (§3.2 #5) $f(x) = \sqrt{x+1}$.

   b) (§3.2 #9) $f(x) = ax^2 + b$ (a, b constants).

   c) $f(x) = x^7$.

   (You will need to use your knowledge of algebra to help *Mathematica* do some of the simplification. In part c) the **Expand** command is useful.)

§2 Computing Derivatives

We have seen already how to use *Mathematica*'s "prime" notation for derivatives: if **f[x]** is a *Mathematica* function, **f'[x]** is its derivative. We now introduce a second notation: *Mathematica*'s **D** command for computing derivatives.

Example 1 Use the **D** command to find the second derivative of $V(r) = \pi r^2 h$ with respect to r.
Solution:

```
D[Pi r^2 h,{r,2}]
```
The r inside the list bracket tells Mathematica the variable with respect to which the derivative is to be taken; all other variables are treated as constants. The number 2 following the r is the order of the derivative. Note that we didn't need to define $V(r) = \pi r^2 h$ as a Mathematica function — the D command can be applied to many types of expressions.

Example 2 Compute the first four derivatives of the function $f(x) = x^2 \sin(\pi x)$.
Solution:

```
f[x_]  =  x^2  Sin[Pi  x]

D[f[x],x]
D[f[x],{x,1}]
```
For first derivatives the list brackets and the 1 are optional. We could also have dispensed with

defining *f(x)* *as a function — instead just giving the name f to the expression* $x^2 \sin(\pi x)$:

```
Clear[f]
f = x^2 Sin[Pi x]
D[f,x]
D[f,{x,2}]
D[f,{x,3}]
D[f,{x,4}]
```

Exercises

1) Compute the derivative $f'(x)$ by hand and then check by using *Mathematica*.

    a) $f(x) = x \cos x$

    b) $f(x) = \dfrac{x \sin(3x)}{2x + 9}$

    c) $f(x) = 3\tan(k(x))$, where $k(x)$ is an unspecified differentiable function.

2)   Let $f(x) = A\cos(3\pi x)$.

    a) Find the third derivative of f with respect to the variable x.

    b) Find the third derivative of f with respect to the variable A.

§3 Implicit Differentiation

    In this section we introduce a method for plotting graphs of implicitly defined functions (level curves) and then show how *Mathematica* can be used to find derivatives of such functions.

Example 1 Plot the graph of the equation $x^2 y^2 + x \sin y = 1$. Estimate the slope of this graph where $x = 1$.

<u>Solution:</u>

    The **ContourPlot** *command can be used to obtain the graph. This is a command for drawing the contour curves (level curves) of a function of two variables. (See pages 114-116 for more information on this command and its options) By specifying* **Contours** -> {1}, *we* get the level curve $x^2y^2 + x\sin y = 1$. We use the option `ContourShading -> False` so that the plot will not be shaded.

```
ContourPlot[x^2 y^2 + x Sin[y], {x,-5,5},{y,-5,5},
     Contours -> {1},ContourShading -> False]
```

While this plot shows the basic shape of the curve, it is too jagged. We can improve the plot by increasing the value of the option `PlotPoints` and "smoothing" the contour curve:

```
ContourPlot[x^2 y^2 + x Sin[y], {x,-5,5}, {y,-5,5},
     Contours -> {1}, PlotPoints -> 30,
     ContourSmoothing -> Automatic, ContourShading -> False]
```

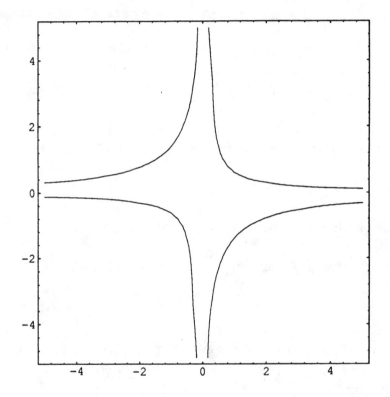

We see that there are two points on the curve with x-coordinate 1. One of these is on the top part of the graph and one on the lower part. (Each part is the graph of a function defined implicitly by the equation $x^2y^2 + x\sin y = 1$.)

Using the cursor crosshairs we see that the two points we want are approximately (1, 0.6) and (1, −1.3). The slopes at these points appear to be about −1 and +1, respectively. (Note that the plot has equal scales on the axes, so a line making a 45° angle with the x-axis has slope 1.)

Example 2 Continuing our analysis of the curve in Example 1, calculate the slopes at the two points on the curve with x-coordinate 1.

Solution:

First we find the y-coordinates of these two points. Substituting $x = 1$ into the equation yields $y^2 + sin(y) = 1$. We use the **FindRoot** command to solve this nonlinear equation for y, using the estimates .6, and −1.3 as starting values. (Unlike **Solve** which algebraically solves an equation, **FindRoot** finds an approximate solution (root) of an equation, starting its search at the given point. The method used is similar to Newton's method, which you will be studying later on in your calculus class.)

```
FindRoot[ y^2 + Sin[y] == 1, {y,0.6}]
FindRoot[ y^2 + Sin[y] == 1, {y,-1.3}]
```

The approximations we get for the two points are (1, 0.636733) and (1, −1.40962). Next we find the slope of the curve at a typical point (x, y) by implicit differentiation:

```
D[ x^2 y[x]^2 + x Sin[y[x]] == 1, x]
```

(Note that we applied the D command to an equation, and the output is an equation. Also , since y is a function of x, it is necassary to replace y by y[x] so that the dependence of y on x is made explicit.)

We solve this equation for y'[x]:
```
Solve[%,y'[x]]
```

This gives the formula for the derivative $y'[x] = \dfrac{dy}{dx}$, which is also the slope at an arbitrary point (x, y). Write this as a function of both x and y:
```
slope[x_,y_] = -(Sin[y] + 2x y^2) / (x Cos[y] + 2x^2 y)
```

Then evaluate the derivative at our two points of interest:
```
slope[1,  0.636733]
```

```
slope[1,  -1.40962]
```

Compare these more accurate approximate values with the values estimated from the graph in Example 1: the slope at the upper point is about −0.636 which is in reasonable agreement with our crude estimate of −1 for this slope. Likewise the slope at the lower point is about 1.12, so our estimate of 1 was fairly accurate.

33

Example 3 Find the second derivative $\dfrac{d^2y}{dx^2}$ at a typical point (x, y) on the curve $x^2y^2 + x\sin y = 1$ studied in Examples 1 and 2 above.

<u>Solution:</u>

Our first step is to differentiate the expression for $\dfrac{dy}{dx}$ *found in Example 2 , again using implicit differentiation. The resulting expression for* $\dfrac{d^2y}{dx^2}$ *will involve* y'[x].

```
d2  =  D[slope[x,y[x]],  x]
```

Now just replace y[x] *everywhere by* **y** *and substitute the expression* **slope[x,y]** *in place of* y'[x], *to get the desired formula for* $\dfrac{d^2y}{dx^2}$ *in terms of just x and y.*

```
ReplaceAll[d2,  {y[x]-> y,  y'[x]-> slope[x,y]}].
```

This is a mess, so we simplify it:
```
d2  =  Simplify[%]
```

Exercises

For each of the following problems:

i) Plot the graph of the equation and estimate the slope at the given point on the graph.

ii) Use *Mathematica* to help calculate the derivative $\dfrac{dy}{dx}$ at the given point. Does the result agree with the slope estimated in part (i)?

iii) Use *Mathematica* to help calculate the second derivative $\dfrac{d^2y}{dx^2}$ at the given point. What does your answer say about the curve at the given point?

1) (§3.6 #30) $x^2y - 5xy^2 + 6 = 0$; (3,1).

2) $x^3 + y^3 = 3xy$; $(\tfrac{3}{2}, \tfrac{3}{2})$.

3) (§3.6 #32) $\sin(xy) = y$; $(\tfrac{\pi}{2}, 1)$.

Chapter 4 - Applications of the Derivative

New *Mathematica* Commands: NestList

§1 Intervals of Increase, Decrease and Concavity

In this section we will investigate how the signs of the first two derivatives of a function are related to the geometry of the graph of that function.

Example 1 Plot the graph of the function $f(x) = (x^4 - 30x^2 - 60)\cos(x)$, along with its first and second derivatives, over the interval $[-2\pi, 2\pi]$.

a) Over what intervals does f appear to be increasing?

b) Over what intervals does f' appear to be positive?

c) Compare your answers to parts a) and b).

d) Over what intervals does the graph of f appear to be concave up?

e) Over what intervals does the second derivative f'' appear to be positive? Compare your answer to the answer to part d). Is every point where f'' is zero a point of inflection of the graph of f?

<u>Solution:</u>

Begin by defining the function f and plotting its graph and that of its derivative:
```
f[x_]  =  (x^4  -  30x^2  -  60)  Cos[x]

Plot[{f[x],f'[x],f''[x]},{x,-2Pi,2Pi},PlotStyle->
    {RGBColor[0,0,0],RGBColor[1,0,0],RGBColor[0,0,1]}]
```

a) The function is increasing on the intervals (−5.2, −3.3), (0, 3.3), (5.2, 2π).

b) The derivative is positive on the intervals (−5.2, −3.3), (0, 3.3), (5.2, 2π).

c) The answers in a and b are the same. We expected this since a positive derivative implies an increasing function.

d & e) The graph of f appears to be concave up on the intervals $(-2\pi, -4.4)$, $(-2.3, 2.3)$ *and* $(4.4, 2\pi)$, *and these are exactly the intervals on which the second derivative appears to be non-negative. The second derivative appears to be zero at* $x = 0$, *and this is reflected in the fact that the graph of f is nearly straight there, i.e., its slope is nearly constant near* $x = 0$. *Note that* $(0, -60)$ *is not a point of inflection , even though* $f''(0) = 0$.

Example 2 Plot the graph of the function $f(x) = 4x^5 - 5x^3$. Choose a domain for your plot which will include all critical points of f and all inflection points of the graph of f.

Solution:

```
f[x_]  =  4x^5  -  5x^3
```

The critical points of f are the x-values in the domain of f at which either f' *is zero or* f' *fails to exist. In this example the function f is differentiable everywhere, so we can find the critical points by solving the equation* $f'(x) = 0$.

```
Solve[f'[x]  ==  0,x]
```

Note that the critical point $x = 0$ *is neither a local maximum nor a local minimum of f.*

The points of inflection are points at which the second derivative changes sign. Since in this example f'' *is continuous, it can change sign only by passing through zero. Thus the points of inflection will be among the solutions of the equation* $f''(x) = 0$:

```
Solve[f''[x]  ==  0,x]
```

All the critical points and roots of the second derivative are between -1 *and* 1, *so we choose this interval as the domain for our Plot command:*

```
Plot[f[x],{x,-1,1}]
```

From the graph we note that all three solutions of $f''(x) = 0$ *correspond to points of inflection of the graph of f.*

Example 3 If $f(x) = \dfrac{x^3 - 1}{x^3 + 1}$, use *Mathematica* to find all critical points of f, and all inflection points and all horizontal or vertical asymptotes of the graph of f.

Solution:
Define the function and make a preliminary plot of its graph:

```
f[x_]  =  (x^3  -  1)/(x^3  +  1)
```

36

```
Plot[f[x],   {x,-3,3}]
```

To find the critical points we calculate the derivative and simplify it:
```
Simplify[f'[x]]
```

We see that the derivative is zero only at zero. We also note that the derivative does not exist at $x = -1$. However $f(-1)$ is undefined, so $x = -1$ is not a critical point for the function. Thus the only critical point is $x = 0$. From the plot it is clear that the line $x = -1$ is a vertical asymptote of the graph: $\lim_{x \to -1^-} f(x) = \infty$ *and* $\lim_{x \to -1^+} f(x) = -\infty$.

Find the inflection points:
```
Simplify[f''[x]]
```

We see that f'' is continuous for all $x \neq -1$, so the only way it can change sign on $(-\infty, -1)$ or $(1, \infty)$ is by passing through zero. Thus the points of inflection are among the roots of f'' :
```
Solve[Numerator[%]   ==   0,x]
N[%]
```
The only real solutions are $\dfrac{1}{\sqrt[3]{2}}$ and 0.

We plot the graph of the second derivative to see if these correspond to points of inflection of the graph of f:
```
Plot[f''[x],{x,-2,2},PlotRange->{-5,5}]
```

We see that f is concave up on $(-\infty, -1)$ and on $(0, \dfrac{1}{\sqrt[3]{2}})$. It is concave down on $(-1, 0)$ and on $(\dfrac{1}{\sqrt[3]{2}}, \infty)$. So the points of inflection occur where $x = 0$ and $x = \dfrac{1}{\sqrt[3]{2}}$.
```
f[0]
f[1/2^(1/3)]
```
Thus $(0, -1)$ and $(\dfrac{1}{\sqrt[3]{2}}, -\frac{1}{3})$ are the points of inflection of the graph of f.

Finally, does the graph of f have a horizontal asymptote?

```
Limit[f[x],   x->Infinity]
Limit[f[x],   x->-Infinity]
```
The line $y = 1$ is a horizontal asymptote, approached by the graph of f both as $x \to -\infty$ and as $x \to \infty$.

We can check our results by plotting y = f(x) and the horizontal asymptote y = 1:

```
Plot[{{f[x],1},{x,-3,3},PlotRange->{-5,5},
    PlotStyle->{RGBColor[0,0,0],RGBColor[0,0,1]}}]
```

Remark In the solution to Exercise 3 above we glossed over a problem with *Mathematica*. The command **Solve[Numerator[%] == 0, x]** produced the output

$$\{\{x \to 2^{-(1/3)}\}, \quad \{x \to \frac{(-1)^{2/3}}{2^{1/3}}\}, \quad \{x \to \frac{(-1)^{4/3}}{2^{1/3}}\}, \quad \{x \to 0\}\}$$

which indicates four real number solutions to the equation. However, the next command **N[%]** showed that the second and third solutions in this list are complex numbers. In order to clear up the confusion here, we need to digress a bit to explain cube roots of numbers in a little more detail than is customary in introductory calculus.

Every nonzero complex number has in the complex number system two distinct square roots, three distinct cube roots, four distinct fourth roots, etc.. In elementary calculus we want to avoid dealing with complex numbers, so *we simply ignore all the roots of a real number which are not real numbers themselves.* Thus we say that a positive real number has two distinct square roots but negative real numbers have no square roots. Similarly we say a positive real number has only one cube root, and also a negative real number has a unique cube root. For example the cube root of 8 is 2 and the cube root of –8 is –2. *Mathematica*, however, is a system designed for doing advanced mathematics involving both real and complex numbers, and in such work our preference for real roots is too restrictive. It turns out that if c is a positive real number, *Mathematica* denotes by $c^{\frac{1}{3}}$ the positive real cube root of c, as we do in calculus. For example, the value of $8^{\frac{1}{3}}$ is 2. However if c is negative, the *Mathematica* command **c^(1/3)** produces the complex number $|c|^{\frac{1}{3}}(-1)^{\frac{1}{3}}$, where $(-1)^{\frac{1}{3}}$ is the complex number $\frac{1}{2} + \frac{\sqrt{3}}{2}I$. For example *Mathematica*'s value for $(-8)^{\frac{1}{3}} = 2(-1)^{\frac{1}{3}}$ is $1 + \sqrt{3}I$, rather than –2. Both answers are correct, but in introductory calculus we want the cube root of –8 to be –2. How can we persuade *Mathematica* to give us the real cube root of negative real numbers? The simplest way is to get inside the *Mathematica* program and give it a new rule for calculating powers. Enter

```
Unprotect[Power]
Power[x_,Rational[n_,3]]  :=  (- Abs[x]^(1/3))^n  /;  x < 0
Protect[Power]
```

This tells *Mathematica* to use the rule $x^{\frac{n}{3}} = (-|x|^{\frac{1}{3}})^n$ whenever x is negative. This rule will be in effect for the remainder of the current *Mathematica* session. To check that this rule has the desired effect, try evaluating `(-8)^(1/3)` and `(-8)^(2/3)` to make sure you get –2 and 4.

We need this rule for computing cube roots of negative numbers in the following example.

Example 4 Plot the graph of the function $g(x) = 2x^{\frac{1}{3}} - x^{\frac{2}{3}}$. Determine all critical points of g, and classify these points. Also find any points of inflection and all vertical or horizontal asymptotes of the graph.

Solution:

First define the function and make a preliminary plot:
```
g[x_]  =  2x^(1/3)  -  x^(2/3)
```

We assume that Mathematica's **Power** *function has been modified as indicated above, before entering the* **Plot** *command, so that g(x) will be properly interpreted for negative values of x.*
```
Plot[g[x],   {x,-3,3}]
```

There appears to be a vertical tangent line at the origin, and a local maximum near $x = 1$. A graph of the first derivative may clarify this:
```
Plot[g'[x],   {x,-3,3}]
```

We see that indeed the slope of the graph approaches infinity as $x \to 0$, so the tangent line at the origin is vertical, and we see that the slope changes from positive to negative near $x = 1$, so this point is a local maximum. To be certain that the x-value at this critical point is exactly 1 and to make sure there are no other critical points, we examine the formula for the first derivative:
```
g'[x]
```

```
Simplify[%]
```
The result is $\dfrac{2(1-x^{\frac{1}{3}})}{3x^{\frac{2}{3}}}$, *which approaches infinity as $x \to 0$ and is zero only at $x = 1$. This*

39

confirms our graphical analysis.

Our plot of the graph of g shows that the origin is a point of inflection. To see if there are any other inflection points, we examine the second derivative:
`Simplify[g''[x]]`

The result, $\dfrac{2\,(-2+x^{\frac{1}{3}})}{9\,x^{\frac{5}{3}}}$, *changes from positive to negative as x passes through* 0, *and changes from negative to positive as x passes through* 8. *Thus besides the point of inflection at the origin there is another, at* (8, 0).

Since g is continuous for all x, the graph has no vertical asymptotes. Also for large values of $|x|$ *the term* $x^{\frac{2}{3}}$ *is much larger than* $2x^{\frac{1}{3}}$, *so* $\lim\limits_{x\to\infty} g(x) = \lim\limits_{x\to-\infty} g(x) = -\infty$. *Thus the graph has no horizontal asymptotes.*

Exercises

1) For the functions below find all intervals on which f is
   a) increasing,   b) concave up, and   c) find all points of inflection.
   Also compare the graph of each function to the graph of its derivative, as in Example 1.
   i)   (§4.2, #6) $f(x) = 4 - 3x - x^2$
   ii) (§4.2, #8) $f(x) = 5 + 12x - x^3$
   iii) (§4.2, #18) $f(x) = x^{\frac{2}{3}} - x^{\frac{1}{3}}$ (Be sure *Mathematica*'s **Power** function has been modified as indicated in the Remark preceding Example 4 above, before starting this problem.)

2) For each of the following functions make a plot which shows as many interesting features of the graph of the function as possible.
   i) (§4.5, #10) $f(x) = \sqrt[3]{x^2 - 4}$ (Be sure *Mathematica*'s **Power** function has been modified as indicated in the Remark preceding Example 4 above, before starting this problem.)

   ii) (§4.5, #16) $f(x) = \dfrac{1 + \sqrt{x}}{1 - \sqrt{x}}$

§2 Applied Maximum and Minimum Problems

This section shows how *Mathematica* can be an aid in the solution of applied optimization problems.

Example 1 Find the point on the graph of $y = 1 + 4\sin 2x$ which is nearest to the point (1, 1). What is the minimum distance?

Solution:

Plot the graph of the function and estimate the nearest point to (1, 1):
```
f[x_]  =  1  +  4Sin[2x]
```

```
Plot[f[x],    {x,0,2}]
```
Using the crosshairs we estimate a nearest point of (1.5, 1.8).

To get a better estimate, define a function which gives the distance to the point (1, 1) *from any point* (x, f(x)) *on the graph of f .*
```
d[x_]  =  Sqrt[(x  -  1)^2  +  (f[x]  -  1)^2]
```

Plot the graph of this function:
```
Plot[d[x],    {x,0,4}]
```

From this plot we estimate that the x-coordinate of the nearest point is about 1.56, *and the distance from this point to* (1, 1) *is about* 0.56. *To get an even more accurate estimate we use the fact that the minimum value of d[x] occurs at a critical point of the function d, i.e., a root of the derivative d'[x]. We have an approximate root* $x \approx 1.56$, *so we can use the FindRoot command:*

```
FindRoot[d'[x]   ==   0,   {x,1.56}]
xmin  =  ReplaceAll[x,%]
```
Even though there are an infinite number of solutions to this equation, our graphical analysis shows that the solution near 1.56 *gives the x-coordinate of the point on the graph of f with minimum distance from* (1, 1).

What is the minimal distance and for which point does it occur?

```
y  =  f[xmin]
```
The point on the graph of $y = 1 + 4\sin 2x$, *nearest to* (1, 1) *is approximately* (1.56201, 1.07026).

41

d[xmin]
The distance from (1.56201, 1.07026) *to* (1, 1) *is about* 0.566388.

Exercises

1. (§4.7, #33) Prove that (1, 0) is the closest point on the curve
   $x^2 + y^2 = 1$ to (2, 0).

2. (§4.8, #26) A ladder is used to reach over a fence 8 feet high to rest against a wall that is 1 foot behind the fence. What is the length of the shortest ladder that can be used? (Hint: Express the length of the ladder as a function of its angle of elevation.)

3. A cockroach and a spider move in the xy plane, the location of the roach at time t being given by $\begin{cases} x_1(t) = 2\cos t \\ y_1(t) = 3\sin t \end{cases}$ and the location of the spider being
$\begin{cases} x_2(t) = 5\cos 2t \\ y_2(t) = 3\sin 2t \end{cases}$. At what instant in the time interval
$0 \le t \le 3$ is the distance between the roach and spider minimized? What is the minimal distance?

§3 Newton's Method

Because it is highly computational Newton's method is best used with the aid of a computer. *Mathematica's* **FindRoot** command uses Newton's method to find a very accurate approximation to a root of a function, starting from a crude approximation to the root. But **FindRoot** just gives as output the final approximation, hiding the intermediate steps. In this section we will see in detail the sequence of approximations produced by Newton's method.

Example 1 Use Newton's method to find a root of $f(x) = x^3 - x - 1$.

Solution:

We begin by defining the function and plotting its graph:
```
f[x_] = x^3 - x - 1
Plot[f[x],{x,-3,3}]
```
From the graph it is clear that our equation has only one solution, $x \approx 1.3$.

We define a function called **Newton** which will have as input an approximate root of the function f, and will output the next approximate root as given by Newton's method. (In honor of Isaac Newton we will make an exception to the rule that user-defined functions in Mathematica should have names beginning with a lower-case letter.):

```
Newton[x_] = x - f[x]/f'[x]
```

Try this function using an initial approximation of 2:

```
Newton[2]
```
The output is in fractional form since Mathematica treats 2 as an exact rational number; we would prefer decimal values, so we change 2 into a 20-digit decimal number:

```
Newton[N[2,20]]
```
We repeat this procedure, making each output in turn an input until the output stabilizes:
```
Newton[%]
Newton[%]
Newton[%]
Newton[%]
Newton[%]
Newton[%]
```
Since the number has been repeated once, it will now just continue to repeat forever. We have as accurate an answer as we can obtain within our precision. The root is approximately $x = 1.324717957244746026$. One test of an approximate root is to see how close to zero the function is at the point:
```
f[%]
```
This is a very small number. Apparently our approximation is a good one.

Example 2 Use Newton's method to find a root to $g(x) = x^5 + 2x - 10$.

Solution:

```
g[x_] = x^5 + 2x - 10
Plot[g[x],{x,-3,3}]
```
There is only one root, $x \approx 1.5$:
We define the Newton function corresponding to g as before:

```
Newton[x_] = N[x - g[x]/g'[x], 20]
```

43

Rather than repeating the same command over and over as in the previous example we will let the computer do the work for us. For any Mathematica function p and any number x0, the command **NestList[p, x0, n]** will create the list consisting of x0, p[x0], p[p[x0]], p[p[p[x0]]], p[p[p[p[x0]]]], ..., until the function p has been iterated n times. We apply this to our Newton function:

```
NestList[Newton,  N[1.5,20],  6]
```

The values stabilize quickly, but Mathematica has displayed only the first few digits of the successive approximations to the root. This is because Mathematica considers the starting value 1.5 to be a low-precision approximate number. We can get the full 20-digits in all the successive approximations by replacing our starting value 1.5 by the rational number 3/2:

```
NestList[Newton,  N[3/2,20],  6]
```
Check the final result by substitution:
```
Last[%]
```
This picks out the last entry in our list of approximations to the root of g.

```
g[%]
```
As expected, Newton's method led to the desired root of g.

Exercises

1. Use a plot and Newton's method to find a root of $x^7 - x - 1$. How many real roots does this polynomial have? Verify that **FindRoot** produces the same final result as Newton's method.

2. Use graphical methods to determine how many real solutions the equation $\cos x = x$ has. Then use Newton's method to approximate each of these solutions to 10 digits. (You must write the equation in the form $f(x) = 0$, so that solving the equation is equivalent to finding the roots of the function f.)

3. Solve $x^3 - 3x = 1 - \sin x$.
   a) Find the number of (real) solutions, and approximate values of these solutions, using graphical methods.

   b) Use Newton's method to find each root to 10 digits.

Chapter 5   The Integral

	Sum
New *Mathematica* Commands:	Needs
	Integrate
	NIntegrate

§1 Approximation by Rectangles - Computation

We will show how *Mathematica* can be used to compute various types of Riemann sums.

Example 1 Approximate the area under the graph of the function $f(x) = \dfrac{x^4 - 9}{3x - 8}$ over the interval [-1, 1.5], using 40 rectangles of equal width. Determine the height of your rectangles by using :
     a)   the left endpoint of each interval.
     b)   the right endpoint of each interval.
     c)   the midpoint of each interval.

Solution:

a) *First we define the function and plot its graph over the interval* [-1, 1.5]:

```
f[x_]  =  (x^4  -  9)/(3x  -  8)
```

```
Plot[f[x],{x,-1,1.5},PlotRange  ->  {0,2}]
```
The function f is positive over [-1, 1.5], *so the area under its graph is given by the integral*
$\int_{-1}^{1.5} f(x)\, dx$. *To make our solution quite general, we will let a denote the left endpoint of the interval of integration, b the right endpoint and m the number of rectangles used to approximate the area. Start by entering these commands:*
```
a  =  -1;
b  =  1.5;
m  =  40;
```

We divide the interval [a, b] *into m subintervals of equal width. Enter:*

45

```
dx  =  (b-a)/m
```

The left endpoint of the interval of integration is the first partition point:
x[0] = a. The next partition point will be a distance dx to the right of a:
x[1] = a + 1 dx. In general we have that the n-th partition point is:
x[n] = a + n dx.   So the partition points can be created by entering a single command:

```
x[n_]  =  a + n  dx
```

To approximate the area under the curve we must sum up the areas of m rectangles, each of width dx, whose heights are the values of f at "sample points" chosen from the subintervals of our partition. In part a) these sample points are to be the left endpoints of the subintervals of our partition.

The left endpoint of the first subinterval is	x[0].
The left endpoint of the second subinterval is	x[1].
The left endpoint of the third subinterval is	x[2].
In general, the left endpoint of the n-th subinterval is	x[n-1].

Thus the height of the n-th rectangle is f[x[n-1]], and so its area is f[x[n-1]] dx. Adding these together gives the desired Riemann sum approximation to the area. Enter

```
leftSum  =  Sum[  f[x[n-1]]  dx, {n,1,m}]
```

We conclude that the area under the curve is approximately 3.03832.

Mathematica note:

We have used a new command: `Sum[expr[n],{n,1,m}]` finds the sum of the terms expr[1] + expr[2] + expr[3] + … + expr[m].

b) To find the right-hand sum, we need only change the sample points used to find the heights of the rectangles. The left endpoint of the n-th rectangle was x[n-1], the right endpoint is x[n]:

```
rightSum  =  Sum[  f[x[n]]  dx, {n,1,m}]
```
This gives the approximation 3.06318, for the area under our curve.

c) Since the midpoint of the n-th subinterval will be   (x[n] + x[n-1])/2 , enter:

```
midpointSum  = Sum[  f[(x[n-1] + x[n])/2]  dx, {n,1,m}]
```

This approximation is 3.05243.

Example 2 Repeat Example 1, only this time use circumscribed rectangles.

<u>Solution:</u>

With circumscribed rectangles we need to evaluate f at the highest point within each subinterval. When f is increasing on a subinterval, this means we evaluate the function at the right-hand endpoint. For f decreasing, we evaluate at the left-hand endpoint. If f is neither increasing nor decreasing throughout a subinterval then we must work harder to find the maximum value of f on that subinterval.

From the graph it appears that f increases to a maximum near x = 1, then decreases. To check this we set $f'(x) = 0$:
```
Solve[f'[x]==0,x]
```
Wow! Want to try this by hand? We will switch to a numerical approximation:

```
N[%]
```

Two of these solutions are real, and only one, x = 1.06383, lies within the interval [–1, 1.5]. Thus f is increasing on (-1, 1.06383) and decreasing on (1.06383, 1.5).

Now which of the subintervals of our partition lie inside (-1, 1.06383)? Because $x[n] = a + n\,dx \leq 1.06383$ just if $n \leq \dfrac{1.06383 - a}{dx} \approx 33.02$, on the first 33 subintervals of our partition the function f is increasing. On the 34th subinterval, $[x[33], x[34]] = [1.0625, 1.125]$, f is neither increasing nor decreasing; its maximum value on this interval occurs at 1.06383. On the remaining subintervals f is decreasing. Thus we have three groups of subintervals to deal with:

i) On the first 33 subintervals we want to evaluate f at the right endpoint of the subintervals. The height of the n-th rectangle will therefore be `f[x[n]]`, *for* `n = 1` *to 33.*

ii) On the 34th subinterval the maximum value of f occurs at x = 1.06383..., so the height of the 34th rectangle is `f[1.06383...]`.

iii) On the remaining subintervals f is decreasing, so its maximum occurs at the left endpoint. The height of the n-th rectangle is given by `f[x[n-1]]`, *for* `n = 35` *to 40.*

The corresponding Riemann sum approximation to the area is

```
upperSum  =   Sum[  f[x[n]]   dx,{n,1,33}]+  f[1.06383]  dx
                    +  Sum[ f[x[n-1]]  dx,{n,35,40}]
```

Since we used circumscribed rectangles this approximation, 3.0932, is certain to be larger than the exact area. (Later we'll see how to show that the exact area is about 3.05187. So using the midpoints of the subintervals as our sample points gave our best approximation.)

Exercises

1. Approximate the area under $f(x) = \dfrac{x^4 - 9}{3x - 8}$, over the interval [–1, 1.5], using inscribed rectangles, to get an approximation which is certain to be an underestimate of the area.

2. Approximate the area under the graph of $f(x) = x^7 + 9x^3 + x - 7$, over the interval [1, 2],
a) Using 10 circumscribed rectangles.
b) Using 50 circumscribed rectangles.
c) Using 100 circumscribed rectangles.
d) Repeat, using 10, 50 and 100 inscribed rectangles.
e) Based on your analysis, what would be your best guess at the area? Carefully explain your reasoning.

§2 Integration from the definition

In this section we will use *Mathematica's* **SymbolicSum** package to evaluate some integrals directly from the definition of the integral as a limit of Riemann sums. The first example illustrates the package **SymbolicSum**. In the following examples we will apply it to evaluating some integrals.

Example 1 Use the **SymbolicSum** package to find the following sums:

a) $\displaystyle\sum_{i=1}^{m} i^2$ b) $\displaystyle\sum_{i=1}^{m} \dfrac{4i^5 + 2i^2 + 8}{m^2}$

Solution:

a) *First we need to load the package* **SymbolicSum** *into the computer's memory. To do this we use the command* **Needs** :

Needs["Algebra`SymbolicSum`"]

48

This command calls the package **SymbolicSum** which is inside the packages folder named **Algebra**. *Note that the backquote character ` used in the* **Needs** *command is located just to the left of the spacebar on the keyboard.*

A command of the form **SymbolicSum[f[i],{i,a,b}]** *attempts to find a formula for* $\sum_{i=a}^{b} f(i)$, *where the endpoints of the interval of summation a and b may be variable. (In our examples the upper endpoint will be the variable m.)*

```
SymbolicSum[i^2,{i,1,m}]
```

Check this when m = 3.

$1^2 + 2^2 + 3^2 = 1 + 4 + 9 = 14$, *which agrees with*

$$\frac{3(3+1)(2 \cdot 3+1)}{6} = \frac{12 \times 7}{6} = 14.$$

b) **SymbolicSum[(4i^5 +2i^2 +8)/m^2,{i,1,m}]**

Check when m = 2: We can evaluate the output by replacing m by 2:
```
ReplaceAll[%,m->2]
```
This gives $\frac{79}{2}$, *which agrees with the result of the command*
```
Sum[(4i^5 +2i^2 +8)/2^2,{i,1,2}]
```

Example 2 Use the definition of the integral to evaluate $\int_0^5 x^2 dx$.

Solution:

We begin as usual by forming a partition of the interval of integration, using equally spaced points. The only difference is that now the number of subintervals m is left arbitrary.

```
f[x_] = x^2;
a = 0;
b = 5;
dx = (b-a)/m;
x[n_] = a + n dx;
```

49

A Riemann sum of f relative to this partition of [0, 5] *has the form*
$f(z_1)dx + f(z_2)dx + f(z_3)dx + \ldots + f(z_m)dx$, *where* $z_1, z_2, z_3, \ldots, z_m$ *are "sample points",
one sample point chosen from each subinterval of the partition.*

A basic theorem on the existence of the integral asserts that if f is continuous on [a, b], *then the
limit of these Riemann sums as* $m \to \infty$ *is* $\int_a^b f(x)\, dx$, *regardless of what rule is used to choose
the sample points.*
We will choose the right endpoints of the subintervals as sample points, so
$z_n = x[n]$, *for* $n = 1, 2, \ldots m$.

Then we use the **SymbolicSum** *command to evaluate the Riemann sum:*
```
SymbolicSum[f[x[n]]   dx, {n,1,m}]
```

Finally we compute the limit as the number of subintervals $m \to \infty$:
```
Limit[%,m->Infinity]
```

We conclude that $\int_0^5 x^2 dx = \dfrac{125}{3}$.

In the example above we integrated over the interval [0, 5]. A much more
interesting result appears if we integrate over an interval of variable length: [0, x].

Example 3 Use the definition to evaluate $\int_0^x t^2\, dt$. (Note that we use the variable
t as the variable of integration to distinguish it from the endpoint x of the
integration interval.)

Solution:

*We begin as before by constructing a partition of the integration interval into m equal
subintervals, leaving m as a variable:*
```
a = 0;
b = x;
dt = (b-a)/m;
t[n_] = a + n dt;
```

We compute the sum as before:
```
SymbolicSum[f[t[n]]   dt, {n,1,m}]
```

50

Then compute the limit as $m \to \infty$:

```
Limit[%,m->Infinity]
```

Notice that if we take the derivative of the result, $\dfrac{x^3}{3}$, *we get back our original function* $f(x) = x^2$. *Observation of this relationship in general is a major part of the fundamental theorem of calculus: If f is continuous on* [a, b], *then the function* $G(x) = \int_a^x f(t)\, dt$ *is an antiderivative of f on* [a, b].

Exercises

1. Use the definition of the integral to find $\int_2^7 x^3 - 1 \, dx$.

2. Use the definition of the integral to find $\int_1^x 1 - t^3 \, dt$. Check your answer by finding its derivative.

3. Use the definition of the integral to find $\int_x^3 t^2 + t - 1 \, dt$. Check your answer by finding its derivative.

§ 3 The `Integrate` and `NIntegrate` Commands

Mathematica's `Integrate` command is a powerful tool for finding antiderivatives. Both `Integrate` and the numerical integration command `NIntegrate` can be used to evaluate definite integrals.

Example 1

    a) Find an antiderivative of $f(x) = x^3 \sin(2\pi x)$.

    b) Find an antiderivative of $g(x) = bx\sin(cx)$, where *b* and *c* are constants.

Solution:

a) Simply enter:
```
Integrate[ x^3 Sin[2 Pi x], x]
```
The **x** *following the comma is to let Mathematica know the variable of integration.*

51

We can check our antiderivative by differentiating:
```
D[%,x]
```

b) Enter
```
Integrate[b  x  Sin[c  x],  x]
```
All variables other than the variable of integration x are treated as constants.

Example 2 Compute $\int_3^{10}(x^4 - \dfrac{3}{x^2+8x-5})\,dx$.

Solution:

First we check that the function is continuous on [3, 10].
```
N[Solve[x^2 + 8x -5 == 0,x]]
```

Since the denominator has no roots in the given interval, f is continuous on this interval.

To evaluate the integral by using the Fundamental Theorem of Calculus, first find an antiderivative F of f:
```
F[x_] = Integrate[x^4 -3/(x^2 + 8x - 5), x]
```

Now evaluate F at the right endpoint of the interval of integration, and subtract the value of F at the left endpoint.
```
F[10] - F[3]
```

While this is the exact answer, we may want a decimal approximation:
```
N[%]
```

We get an approximate answer of 19951.1

Mathematica can combine the antiderivative and evaluation steps in a single command as follows:
```
Integrate[x^4  -3/(x^2  +8x  -5),{x,3,10}]
```
This is the same result as obtained above.

Example 3 Evaluate $\int_0^1 x\tan(x)\,dx$.

Solution:

First, note that the function $x\tan(x)$ *is continuous on the interval* [0, 1].

The command
```
Integrate[x  Tan[x],{x,0,1}]
```
eventually produces as output just an echo of this input, which means that the **Integrate** *command is unable to find an antiderivative of* $x\tan(x)$.

However, if we then enter **N[%]**, *after about a minute a decimal approximation of the integral, to 6 significant digits, appears. This approximation is determined by a numerical integration algorithm* <u>*which makes no use of an antiderivative of the function*</u> $x\tan(x)$. *(We will study such numerical integration algorithms in a later chapter.) When the* **N[ ]** *command finds that the previous output is* **Integrate[x Tan[x], {x,0,1}]**, *it automatically applies the "numerically integrate" command* **NIntegrate[ ]** *to the function* $x\tan(x)$ *over the interval* [0, 1]. *A much faster and more direct way to reach the same result is to call the* **NIntegrate** *command directly:*
```
NIntegrate[x  Tan[x],  {x,0,1}]
```

Example 4 Find the area between the graph of $f(x) = x^4 - 54x^2 - 40x + 525$ and the x-axis.

Solution:

Define the function and plot it:
```
Clear[f]
f[x_] = x^4 -54 x^2 -40 x + 525
Plot[f[x],{x,-11,10},PlotRange->{-500,1000}]
```
(We added the option **PlotRange->{-500,1000}** *after several attempts to get a nice plot.)*

From the plot we see two regions of interest, one above the axis from about –5 to 3, and one below the axis from 3 to 7. Let's check that these numbers are correct:

```
Solve[f[x]==0,x]
```
So the function is, indeed, zero at –5, 3, and at 7.

```
area1  =      Integrate[f[x],{x,-5,3}]
area2  =  -   Integrate[f[x],{x,3,7}]
```

Then to obtain the total area we compute:

```
area = area1 + area2
```

Exercises

1. (§5.8, #20) Find an antiderivative of $f(x) = \sqrt{\tan x}\sec^2 x$, and then use this to find $\int_0^{\pi/4} \sqrt{\tan x}\sec^2 x\,dx$.

2. (Supplementary Exercises for Chapter 5, #18)
   Compute $\int_{-2}^2 f(x)\,dx$, where $f(x) = \begin{cases} x^3 \text{ for } x \geq 0 \\ -x \text{ for } x < 0 \end{cases}$.

3. Find $\int_1^5 |x-2|\,dx$, first by using the **Integrate** command, and then by using the **NIntegrate** command.

4. Compute $\int_2^7 f'(x)\,dx$, where $f(x) = \dfrac{\sin^2 x + 5x}{3x^3 - \sqrt{x^4 - 1}}$.

5. (§5.7, #51) Find the total area that is between the curve
   $y = x^2 - 3x - 10$ and the interval [–3, 8].

§4 Calculating the Area Between Two Curves

Example Find the area of the regions between the two curves
$f(x) = x^4 - 3x^2 + 2$ and $g(x) = x^3 + 7x^2 + x - 7$.

<u>Solution:</u>

Define and plot the two functions:
```
f[x_] = x^4 - 3x^2 + 2
g[x_] = x^3 + 7x^2 + x - 7
```

```
Plot[{f[x],g[x]},{x,-5,7},
     PlotStyle->{RGBColor[1,0,0],RGBColor[0,0,1]}]
```
The plot shows three regions, one over each of the following intervals (the endpoints are only approximate): (–2.5, –1.5), (–1.5, 1.5), and (1.5, 3).

Solve to find more precisely where the two graphs cross:

```
Solve[f[x]  ==  g[x],x]
```

```
N[%]
```
So the regions are above the intervals: (−2.36735, −1.16634), (−1.16634, 0.896056), *and* (0.896056, 3.63764).

Now integrate, in each case, the top curve minus the bottom curve:

```
area1=  NIntegrate[g[x]  -  f[x],{x,-2.36735,-1.16634}]
area2=  NIntegrate[f[x]  -  g[x],{x,-1.16634,0.896056}]
area3=  NIntegrate[g[x]  -  f[x],{x,0.896056,  3.63764}]
```

These give the areas of the three regions. Add these three numbers together to get the total area.

```
area = area1 + area2 + area3
```

Exercises

Plot the region(s) between the two curves and find the enclosed area:

1. (§6.1, #8) $y = x^3$, $y = x$

2. (§6.1, #24) $y = x^3 - 2x^2$, $y = 2x^2 - 3x$

3. (§6.1, #14) $x = \sin y$, $x = 0$, from $y = \pi/4$, to $y = 3\pi/4$.

Chapter 6 - Logarithms and Exponentials

```
                                   Log
New Mathematica Commands:
                                   Exp
```

§1 Inverse Functions

Example 1 Determine whether each function is one-to-one on its domain, and if it is, find a formula for the inverse function.

a) (§7.1, #6) $f(x) = x^3 - 3x + 2$

b) (§7.1, #18) $g(x) = \dfrac{2x+1}{x-1}$

c) (§7.1, #22) $h(x) = \dfrac{5}{x^2+1}, \quad x \geq 0.$

Solution:

a) *Define the function and plot its graph:*
```
f[x_]  =  x^3  -  3x  +2
```

```
Plot[f[x],{x,-10,10}]
```
We see from the plot that f is not one-to-one.

b) *Again plot:*
```
g[x_]  =   (2x+1)/(x-1)
```

```
Plot[g[x],{x,-10,10}]
```
The function appears to be one-to-one. (The vertical line x = 1 is not part of the graph of g. However, this line is a vertical asymptote of the graph, so having this asymptote shown in the graph is useful.)

To check that the function is one-to-one we take the derivative:
```
g'[x]
```

```
Simplify[%]
```

g' is always negative, so *g* is one-to-one on the intervals $(-\infty, 1)$ and $(1, \infty)$. *It remains to check that the ranges of the function on these two intervals do not overlap. Writing the formula defining g(x) in the form* $g(x) = \dfrac{2 + \frac{1}{x}}{1 - \frac{1}{x}}$, *which is valid for all* $x \neq 0$, *we see that* $\lim\limits_{x \to \infty} g(x) = 2$ *and* $\lim\limits_{x \to -\infty} g(x) = 2$, *so the line* $y = 2$ *is a horizontal asymptote of the graph of g. Also we see that for large positive values of x the numerator* $2 + \frac{1}{x}$ *is a bit larger than 2 while the denominator* $1 - \frac{1}{x}$ *is a bit smaller than 1, so the value of g(x) is always larger than 2 for large x. Similarly we see that* $g(x) < 2$ *for x large in absolute value but negative, so the two ranges do not overlap. This can be verified in a plot showing the graph of g and the line* $y = 2$:

```
Plot[{g[x], 2},{x, -5, 5},
          PlotStyle->{RGBColor[0,0,0], RGBColor[0,0,1]}]
```

Thus g is one-to one on its domain, and the range of g is $(-\infty, 2) \cup (2, \infty)$.

Next compute $g^{-1}(y)$, *for any* $y \neq 1$. *To do this, solve* $y = g(x)$ *for x:*

```
Solve[y == g[x],x]
```

We conclude that $g^{-1}(y) = \dfrac{y+1}{y-2}$.

```
invg[x_] = (x+1)/(x-2)
```

We check that g composed with g^{-1} *at y is y, for any* $y \neq 2$:

```
g[invg[y]]
Simplify[%]
```

and that g^{-1} *composed with g at x is x, for any* $x \neq 1$:

```
invg[g[x]]
Simplify[%]
```

c) *Begin as above:*

```
h[x_] = 5/(x^2+1)
```

Remember that the domain of h was defined to be $x \geq 0$, *rather than the largest possible set for which the formula makes sense.*

```
Plot[h[x],{x,0,10}]
```

From this plot it appears that h is one-to-one.

Check the derivative:

```
h'[x]
```

h is decreasing for $x \geq 0$, *and so h is one-to-one.*

We solve for h^{-1} :

```
Solve[y == h[x],x]
```

Since we required that $x \geq 0$ we need to take the positive square root:

```
invh[x_] = Sqrt[5/x-1]
```

Check:

```
invh[h[x]]
h[invh[y]]
```

While we have finished the problem given, we will plot the graphs of the function, its inverse, the line $y = x$, and check the symmetry:

```
Plot[ {h[x],invh[x],x }, {x,0,10},  PlotStyle->
      {RGBColor[0,0,0],RGBColor[0,0,1],RGBColor[0,1,0]},
      PlotRange->{0,10}]
```

Exercises

Determine whether each function is one-to-one on its domain, and if it is, find a formula for the inverse function.

1. (§7.1, #8) $f(x) = x^3 - 3x^2 + 3x - 1$
2. (§7.1, #20) $f(x) = \sqrt[5]{4x + 2}$
3. (§7.1, #38) $f(x) = 3x^2 + 5x - 2, \quad x \geq 0$

§2 The natural logarithm and exponential functions

The natural logarithm function, sometimes denoted *ln(x)*, is denoted `Log[x]` in *Mathematica*. The logarithm to the base *b* is denoted `Log[b,x]`. The natural exponential, usually written e^x, is denoted `E^x`, or alternatively `Exp[x]`.

Example 1 Plot the function $y = 3 \ln 2x$.

Solution:

Since the logarithm has domain consisting of positive real numbers we plot for positive x only:

```
Plot[3  Log[2x],{x,.01,10}]
```

58

Example 2 Plot the graph of the exponential growth function $y = 100\, e^{.01t}$ and determine its doubling time.
<u>Solution:</u>

```
f[t_] = 100 Exp[.01t]
```

a) *We first investigate the doubling time graphically:*
```
Plot[f[t],{t,0,500}]
```

Using the crosshairs feature we estimate the time it takes to double from 2000 to 4000:

$f(t)$	2000	4000	
t	300	375	*elapsed time:* 75

Similarly we estimate the time required to double from 3000 to 6000:

$f(t)$	3000	6000	
t	340	410	*elapsed time:* 70

So the doubling time appears to be around 70 to 75 time units.

b) *We now find the doubling time analytically:*
For what t does $f(t) = 2000$?
```
FindRoot[f[t] == 2000,{t,150}]
```

For what t does $f(t) = 4000$?
```
FindRoot[f[t] == 4000,{t,300}]
```
If we subtract these values we obtain the doubling time: about 69.315

For a purely algebraic solution, we solve the equation $200 = 100\, e^{.01t}$ for the doubling time t. You should do this by hand, although one could use Mathematica's **Solve** *command. The result is that the doubling time is $t = 100 \ln 2$.*

```
N[100 Log[2],10]
```
This gives the doubling time to 10 decimal places.

Example 3 Show that the exponential function $f(x) = e^x$ grows faster as $x \to \infty$ than does any power function $g(x) = x^n$.
<u>Solution:</u>

We first investigate graphically:

```
Plot[  {Exp[x],x^5},{x,1,20},
            PlotStyle->{RGBColor[1,0,0],RGBColor[0,0,1]}]
```
We see that the exponential function eventually grows larger than the fifth power. Another way to see this is to plot the quotient of the two functions:

```
Plot[  {Exp[x]/x^5},{x,1,20}]
```
Where the exponential function is larger than the power function the quotient is larger than one. Our graph shows that the quotient is very large, for large x. Thus the exponential function grows faster than the fifth power of x.

Try a higher power of x:

```
Plot[Exp[x]/x^12,{x,1,20}]
```
This appears to be going to zero (which would imply that x^{12} is growing faster than Exp[x]) but when we plot on a larger interval we reach a different conclusion:

```
Plot[Exp[x]/x^12,{x,1,100}]
```
Now we see that the exponential function eventually grows much larger than x^{12}. Our graphical investigation indicates that for sufficiently large x the exponential function will be larger than any given power of x. But the graphical approach requires that we look at a specific value of the exponent n in the power function. How can we deal with an arbitrary value of n?

The behaviour of the quotient $\frac{e^x}{x^n}$ for large values of x is described by the limit $\lim\limits_{x\to\infty} \frac{e^x}{x^n}$:

```
Limit[Exp[x]/x^2,x->Infinity]
```
We get an answer of infinity, showing that the exponential function is faster growing than x^2. But when we try a higher power, Mathematica's **Limit** *command is stymied:*

```
Limit[Exp[x]/x^5,x->Infinity]
```

Here a knowledge of logarithms can be applied: applying the natural logarithm (an increasing function) to both sides of the inequality $e^x > x^5$ yields an equivalent inequality $x > 5 \ln x$. Is this last inequality true for large x? We calculate $\lim\limits_{x\to\infty} \frac{x}{5\ln x} = \frac{1}{5} \lim\limits_{x\to\infty} \frac{x}{\ln x}$ and then evaluate this limit:

```
Limit[x/Log[x],x->Infinity]
```
We conclude that e^x grows faster than does x^5. Our argument easily generalizes to show that e^x grows faster than x^n for any positive n, because $\lim\limits_{x\to\infty} \frac{x}{n\ln x} = \frac{1}{n} \lim\limits_{x\to\infty} \frac{x}{\ln x} = \infty$.

Exercises

1. Graphically and analytically find the half-life for the exponential decay function $y = 100\, e^{-.025t}$.

2. Find the doubling time for $y = (500)\, 2^{t/4}$.

3. Suppose a small quantity of a radioisotope which has a half-life of 15.2 days is found to be present in a water sample. If the resulting radiation level is now 70% above the "safe" level, how long will it be before the radiation level in the sample will drop to the safe level? (We assume that the isotope decays into non-radioactive products.)

4. Suppose $1000 is deposited in a bank that pays 12% annual interest,

i) Calculate the amount which will be in the account after 3 years, if the interest is compounded monthly. (That is, each month 1% of the amount in the account is added to the account.)

ii) How long will it take for the amount in the account to double, if the interest is compounded monthly?

iii) Make a table showing the amount which would be in the account after 3 years, if the interest is compounded n times a year, for n = 1, 10, 100, 1000. (Recall that the *Mathematica* command `Table[f[k], {k, 1, m}]` will make a list of the values `f[1]`, `f[2]`, ... `f[m]`, for any function f of an integer variable k.)

iv) Since $\lim_{k \to \infty} (1 + .12/k)^{kt} = e^{.12t}$, we say that 12% annual interest is <u>compounded continuously</u> if an amount A grows to $Ae^{.12t}$ after t years. Compute the amount in our account, with the $1000 initial balance, after 3 years if the 12% interest is compounded continuously.

v) How long will it take for the amount in the account to double, if the interest is compounded continuously?

5. a) Consider the functions $f(x) = \ln x$ and $g(x) = x^{1/n}$, for any fixed positive

integer n. The limit, as x goes to infinity, of each of these functions is infinity. Which of the two functions grows most rapidly, as $x \rightarrow \infty$?

b) Which of the functions, $f(x) = x^x$ or $g(x) = e^x$, grows faster as $x \rightarrow \infty$?

c) Use what you have learned in example 3, together with parts a) and b) of this exercise, to list the following functions in the order of slowest growing to fastest growing, as $x \rightarrow \infty$: e^x, $\ln x$, x^3, x^x, $\ln x^2$, x^5, and $\sqrt{x}$.

Chapter 7   Techniques of Integration

§1 How to change variables in an integral, using *Mathematica*

Mathematica's `Integrate` command allows us to evaluate most of the integrals which appear in typical calculus textbooks. To find an antiderivative $\int f(x)dx$ simply enter `Integrate[f[x],x]`. To calculate a definite integral $\int_a^b f(x)dx$ exactly, just enter `Integrate[f[x],x,{x,a,b}]`. If the program can find an antiderivative F(x) of the function $f(x)$, it produces F(b) - F(a) as output. By the Fundamental Theorem of Calculus this equals $\int_a^b f(x)dx$. As we shall see, however, the user must be alert because in some cases *Mathematica* needs a little help in finding an antiderivative, and occasionally it produces incorrect results.

Let's begin with an example in which *Mathematica* is not able to find an antiderivative, but an observant user sees that a change of variables will lead to a solution.

Example 1 Find $\int \dfrac{x^3}{\sqrt{1-x^8}}dx$.

Solution:

The command `Integrate[x^3/Sqrt[1-x^8],x]` produces an output statement simply repeating the command. This means that *Mathematica* is unable to find the required anti-derivative.

A moment's reflection suggests that the change of variable $u = x^4$ is called for, but the usual procedure for making substitutions in *Mathematica* expressions does not work because *Mathematica* is unable to recognize that $x^8 = u^2$ or that $x^3\, dx = \frac{1}{4}du$, if $u = x^4$. The correct procedure for making the change of variables is therefore to <u>solve for the original variable x in terms of the new variable u.</u> Here $x = u^{\frac{1}{4}}$, so we enter the commands:

```
substitution  =   {x->u^(1/4),   dx->D[u^(1/4),u]}
```

```
ReplaceAll[x^3  dx/Sqrt[1-x^8],substitution]
```

Note that we included the symbol dx as part of the original integrand in the `ReplaceAll` command, but the corresponding symbol du, does not appear in the output of this command. The output involves only the new variable u. The reason for omitting the symbol du is that the `Integrate` command accepts only the expression whose anti-derivative is desired, without the corresponding differential. Hence we proceed by entering:

`Integrate[%,u]` .

The resulting output is `ArcSin[u]/4`. It remains only to express this in terms of the original variable x, using the command

`ReplaceAll[%,u->x^4]` .

The answer just obtained can be checked by differentiation: `D[%,x]` .

Exercises   Find each of the following antiderivatives. First verify that *Mathematica*'s `Integrate` command is unable to calculate these integrals. You may prefer to do the change of variable computations by hand, using *Mathematica* just to find the integrals which result.

1   $\displaystyle\int \frac{x}{\sqrt{16-x^4}}\,dx$

2. $\displaystyle\int \sqrt{\frac{x}{x^3+1}}\,dx$ (Hint: Let $u = x^{3/2}$. In addition to the `Simplify` command, you'll need to use the command `PowerExpand` (which tells *Mathematica* to apply the rule $(a^r)^s = a^{rs}$), to simplify the resulting function of u before integrating.)

3. $\displaystyle\int \frac{\tan(\frac{\pi}{4}+\frac{x}{2})}{\sec^2(\frac{x}{2})}\,dx$

4. $\displaystyle\int \frac{1}{x}\left(\frac{x}{1+x}\right)^{\frac{1}{3}}\,dx$ (Hint: Let $u = \left(\frac{x+1}{x}\right)^{\frac{1}{3}}$, or $\left(\frac{x}{x+1}\right)^{\frac{1}{3}}$, and again use `PowerExpand` before integrating.)

5. $\displaystyle\int \frac{\sin(2x)}{\sqrt{9-(\cos x)^4}}\,dx$

§2 Integration by Parts, using *Mathematica*

Recall that integration by parts is a consequence of the product rule for differentiation, and states that under the proper hypotheses $\int u\,dv = uv - \int v\,du$. *Mathematica* (Version 2.0) appears to use integration by parts to solve all the integration problems which appear in a typical calculus book as applications of this technique. For this reason the examples and exercises below are of a different sort, showing how integration by parts can be used as a tool to derive and apply "recursion formulas", reducing certain integrals to simpler integrals of a similar form.

Example 1 Evaluate $\int (\log x)^n dx$, where n is a positive integer.

Solution:

The command `Integrate[Log[x]^n,x]` produces as output simply an echo of the input, meaning that the `Integrate` command was unable to find the required antiderivative. The difficulty is that *Mathematica* doesn't know how to handle the variable exponent n. It can find the antiderivative for any specific value of n, but cannot find a general formula valid for all n. The key to finding such a formula is integration by parts. Enter the commands

```
u  =  Log[x]^n
du  =  D[u,x]
dv  =  1
v  =  Integrate[dv,x]
v  du
```

The output should be

```
            -1 + n
-n Log[x]
```

The integration by parts formula $\int u\,dv = uv - \int v\,du$ then tells us that

$$\boxed{\int (\log x)^n dx = x(\log x)^n - n\int (\log x)^{n-1} dx}.$$

This formula can be used repeatedly, each time reducing the exponent inside the integral by one until it is zero. Since $\int (\log x)^0 dx = \int 1 dx = x$, the process will always end with this simple integral. For example,

$$\int (\log x)^3 dx = x(\log x)^3 - 3\left(x(\log x)^2 - 2\left(x(\log x)^1 - x\right)\right), \quad \text{or}$$

65

$\int (\log x)^3 dx = x(\log x)^3 - 3x(\log x)^2 + 6x\log x - 6x$. We could develop a general formula for $\int (\log x)^n dx$ as a sum of $n+1$ terms involving decreasing powers of $\log x$, as in this example, but for most purposes the

recursion formula $\int (\log x)^n dx = x(\log x)^n - n\int (\log x)^{n-1} dx$, for $n \geq 1$, and

end condition $\int (\log x)^0 dx = 1$

are a more useful formulation of the idea.

For example, we can define a special *Mathematica* function, let's call it `integrate`, which will use our recursion formula to find antiderivatives of this special form $\int (\log x)^n dx$. (Note that the name of our function begins with a lower-case **i**, so it will be distinct from *Mathematica*'s built-in `Integrate` command.)

Enter the commands:

```
integrate[Log[x_]^n_Integer:1  ,x_]   :=
     x Log[x]^n  -  n integrate[Log[x]^(n-1),x]  /;  n >= 1


integrate[1,  x_]  := x
```

Remarks:

1. The exponent `n_Integer:1` tells *Mathematica* that our command applies only if the exponent is an integer, and if no exponent is present then *Mathematica* is to set $n = 1$ (because `Log[x]` = `Log[x]^1`).

2. The symbols `/; n >= 1` at the end of the first command tell *Mathematica* that the preceding definition is to be used only if $n \geq 1$. (Thus the `/;` symbol can be read "*whenever*".) Without this restriction, if a negative value for n were entered as the exponent, the command would produce an "infinite loop" since each time the rule was applied the exponent would decrease by one, but would never reach zero.

Now we can use our new `integrate` command to find $\int (\log x)^n dx$ for any specific positive integer value of n. For example:

```
integrate[Log[x]^3,x]
```

```
        3                    2
x Log[x]   - 3 (x Log[x]   - 2 (-x + x Log[x]))
Expand[%]
                          2            3
-6 x + 6 x Log[x] - 3 x Log[x]   +  x Log[x]
```

Example 2 Find a recursion formula to evaluate $\int (\sin x)^n dx$ for any integer $n \geq 2$. Use the recursion formula to extend the **integrate** command defined in Example 1, to handle integrals of this form for any $n \geq 0$.
Solution:

We follow the same basic procedure as in Example 1, using integration by parts to find the recursion formula. Enter these commands:

```
u  =  Sin[x]^(n-1)
du =  D[u,x]
dv =  Sin[x]
v  =  Integrate[dv,x]
v  du
```

The output from the last of these commands should be

```
                 2        -2 + n
-((-1 + n) Cos[x]  Sin[x]       )
```

Thus the integration by parts formula $\int u\,dv = uv - \int v\,du$ implies that

$$\int (\sin x)^n dx = -\cos x\,(\sin x)^{n-1} + (n-1)\int (\cos x)^2 (\sin x)^{n-2} dx.$$

Replacing the factor $(\cos x)^2$ by $1 - (\sin x)^2$ gives

$$\int (\sin x)^n dx = -\cos x\,(\sin x)^{n-1} + (n-1)\int (\sin x)^{n-2} dx - (n-1)\int (\sin x)^n dx.$$ Then moving the

factor $(n-1)\int (\sin x)^n dx$ to the left side of the equation yields

$n\int (\sin x)^n dx = -\cos x\,(\sin x)^{n-1} + (n-1)\int (\sin x)^{n-2} dx$ Dividing this equation by n then produces our recursion formula:

$$\int (\sin x)^n dx = \frac{-\cos x\,(\sin x)^{n-1}}{n} + \frac{n-1}{n}\int (\sin x)^{n-2} dx.$$

We can use this formula repeatedly, each time reducing the exponent in the integrand by 2, so that eventually if n is odd we will reach $\int \sin x\,dx = -\cos x$, and if n is even we will end up with $\int (\sin x)^0 dx = \int 1\,dx = x$. In either case, we reduce the original integral to a known one.

To incorporate this recursion formula into our **integrate** command from Exercise 1, we enter the following command:

```
integrate[Sin[x_]^n_Integer,x_]  :=
    -1/n  Cos[x]  Sin[x]^(n-1)  +
    (n-1)/n  integrate[Sin[x]^(n-2), x]  /; n >= 2
```

67

We also enter the "end condition" command:

```
integrate[Sin[x_],x_]   :=   -Cos[x]
```

(Note that the case n = 0 was already taken care of in Example 1.)

You might then test that `integrate` performs correctly by using it to evaluate $\int (\sin x)^4 dx$ and $\int (\sin x)^5 dx$, and verify that it fails to evaluate $\int (\sin x)^{1/2} dx$.

Remark Working through examples like these or the following exercises will give you a glimpse of the stupendous amount of intellectual effort involved in creating just the single command `Integrate` in the *Mathematica* program.

Exercises In each problem, derive the given reduction formula by integration by parts. (You may prefer to do the computations by hand.) Then extend the `integrate` command constructed in Examples 1 and 2 above so that it will also handle integrals of the given type. Be sure to include appropriate "end condition" definitions so that the recursion will terminate correctly. Check by applying the `integrate` command to a few typical examples of the given type.

1. $\int x^n e^x dx = x^n e^x - n \int x^{n-1} e^x dx$

2. $\int (\sec x)^n dx = \dfrac{1}{n-1} (\sec x)^{n-2} \tan x + \dfrac{n-2}{n-1} \int (\sec x)^{n-2} dx$ (Let $dv = (\sec x)^2$, and in the resulting integral replace $(\tan x)^2$ by $(\sec x)^2 - 1$.)

3. $\int (\tan x)^n dx = \dfrac{1}{n-1} (\tan x)^{n-1} - \int (\tan x)^{n-2} dx$ (Start by replacing $(\tan x)^2$ by $(\sec x)^2 - 1$; then use integration by parts to show that
$\int (\tan x)^{n-2} (\sec x)^2 dx = \dfrac{1}{n-1} (\tan x)^{n-1}$.)

4. $\int \dfrac{1}{(x^2 + a^2)^n} dx = \dfrac{x}{2a^2(n-1)(x^2 + a^2)^{n-1}} + \dfrac{2n-3}{2a^2(n-1)} \int \dfrac{1}{(x^2 + a^2)^{n-1}} dx$

   (First, use integration by parts on $\int \dfrac{1}{(x^2 + a^2)^{n-1}} dx$. In the result, replace x^2 by $(x^2 + a^2) - a^2$ and solve the resulting equation for $\int \dfrac{1}{(x^2 + a^2)^n} dx$.)

§3 An example where *Mathematica's* Integrate command yields an incorrect result.

Without the use of *Mathematica*, the most natural method of evaluating $\int_1^3 \left|\ln\left(\frac{x}{2}\right)\right| dx$, is to split the interval of integration into two subintervals determined by the sign of $\ln\left(\frac{x}{2}\right)$. $\ln\left(\frac{x}{2}\right) > 0$ if and only if $\frac{x}{2} > 1$, i.e., $x > 2$.

So $\int_1^3 \left|\ln\left(\frac{x}{2}\right)\right| dx = \int_1^2 -\ln\left(\frac{x}{2}\right) dx + \int_2^3 \ln\left(\frac{x}{2}\right) dx.$

These two integrals can be evaluated by *Mathematica's* `Integrate` command, yielding the result $\mathrm{Log}\frac{1}{2} + 3\mathrm{Log}\frac{3}{2}$.

An interesting phenomenon occurs if instead we attempt to use *Mathematica* to evaluate $\int_1^3 \left|\ln\left(\frac{x}{2}\right)\right| dx$ without splitting the integral up.

Mathematica is unable to integrate many expressions involving explicit reference to the absolute value function, so in the hope of circumventing this unfortunate fact we will replace `Abs[u]` by the equivalent algebraic expression `Sqrt[u^2]`. Enter the command
`Integrate[Sqrt[Log[x/2]^2],x].`

The output is $\left(x - \dfrac{x}{\mathrm{Log}\left(\frac{x}{2}\right)}\right)\mathrm{Sqrt}\left[\left(\mathrm{Log}\left(\frac{x}{2}\right)\right)^2\right].$

When we check by differentiation the command
`Simplify[D[%,x]]`

yields $\mathrm{Sqrt}\left[\left(\mathrm{Log}\left(\frac{x}{2}\right)\right)^2\right]$, so we have an antiderivative as desired. But when we evaluate the definite integral over the interval [1, 3], using the command
`Integrate[Sqrt[(Log[x/2])^2],   {x,1,3}]`
the result has the numerical value -3.69315. This is absurd! The function being integrated is positive throughout the interval [1, 3], except at $x = 2$ where it is zero, so its integral must be positive.

What is wrong? A plot of the function and its antiderivative points out the difficulty:

```
Plot[{Sqrt[(Log[x/2])^2],
    (x  -  x/Log[x/2])*Sqrt[Log[x/2]^2]},{x,  1,  3},
    PlotStyle  ->  {GrayLevel[0.6],GrayLevel[0]}]
```

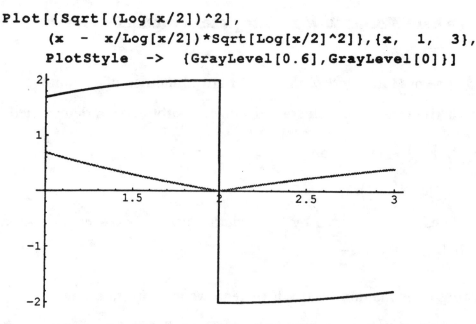

The antiderivative has a discontinuity at $x = 2$. In fact, looking at the formula for the antiderivative, we see that it is undefined at $x = 2$. The Fundamental Theorem of Calculus says that if f is continuous function on $[a, b]$, then $\int_a^b f(x)dx = F(b) - F(a)$ provided that $F'(x) = f(x)$ at <u>every</u> point x of (a, b) and in addition $\lim\limits_{x \to a^+} \dfrac{F(x) - F(a)}{x - a} = f(a)$ and $\lim\limits_{x \to b^-} \dfrac{F(x) - F(b)}{x - b} = f(b)$. This implies that <u>the antiderivative F must be continuous on the closed interval $[a, b]$</u>. So the difficulty is that <u>the alleged antiderivative found by *Mathematica* has a jump discontinuity at one point of the interval of integration</u>.

Our plot of the graphs indicates how to correctly evaluate our integral. We have an antiderivative of $|\log(x/2)| = \sqrt{\log(x/2)^2}$ at every point of $[1, 3]$ except $x = 2$, so the Fundamental Theorem can be applied on every interval $[1, s]$ with $s < 2$ and on every interval $[t, 3]$ with $t > 2$. We break up $\int_1^3 \sqrt{\log(x/2)^2}\,dx$ into the sum $\int_1^2 \sqrt{\log(x/2)^2}\,dx + \int_2^3 \sqrt{\log(x/2)^2}\,dx$ and express these integrals as one-sided limits: $\lim\limits_{s \to 2^-} \int_1^s \sqrt{\log(x/2)^2}\,dx + \lim\limits_{t \to 2^+} \int_t^3 \sqrt{\log(x/2)^2}\,dx$. Therefore the following command will correctly evaluate our integral:

```
exactValue  =  Limit[Integrate[Sqrt[(Log[x/2])^2],
                    {x,1,s}],s-> 2,Direction -> 1]
            +  Limit[Integrate[Sqrt[(Log[x/2])^2],
                    {x,t,3}],t-> 2,Direction -> -1]
```

70

The output is in unevaluated form:

$$4 - (1 - \frac{1}{\text{Log}[\frac{1}{2}]}) \text{ Sqrt}[\text{Log}[\tfrac{1}{2}]^2] + (3 - \frac{3}{\text{Log}[\frac{3}{2}]}) \text{ Sqrt}[\text{Log}[\tfrac{3}{2}]^2]$$

```
N[%,10]
```
0.5232481438
```
N[Log[1/2]  +  3Log[3/2],10]
```
0.5232481438

The results of the two methods agree.

This example should serve as a warning that one must maintain a healthy skepticism of results produced by complex computer algorithms. The results should be checked whenever possible. It also emphasizes the importance of verifying that the hypotheses of a theorem are satisfied before assuming that the conclusion of the theorem is true. The next section discusses some ways one can check the values of definite integrals computed by *Mathematica*'s `Integrate` command.

Exercise Use the `Integrate` command to attempt to evaluate $\int_1^5 |x-2|\, dx$, first expressing the integrand as `Abs[x-2]` and then as `Sqrt[(x-2)^2]`. Check by splitting the integral into a sum of integrals over subintervals of [1, 5]. Which answer is correct? Carefully explain the source of the error.

§4 The Trapezoidal rule and Simpson's rule

The trapezoidal rule provides us with a simple way of obtaining numerical approximations of definite integrals without the need to find an antiderivative. It asserts that under suitable hypotheses $\int_a^b f(x)\, dx$ is approximated by the sum:

$$\frac{h}{2}(f(x_0) + 2f(x_1) + 2f(x_2) + \ldots + 2f(x_{n-1}) + f(x_n)),$$

where $h = \dfrac{b-a}{n}$, $x_0 = a$, $x_1 = a+h$, $x_2 = a+2h$, ..., $x_n = a + nh = b$.

Example 1 Approximate $\int_1^3 e^{\sqrt{x}}\, dx$ using the trapezoidal rule with $n = 10$. Compare this approximation with the estimate obtained using *Mathematica's* numerical integration command: `NIntegrate[E^Sqrt[x], {x,1,3}]`. Repeat, with $n = 20$.

Solution:

Begin by defining the function to be integrated

```
f[x_] = E^Sqrt[x]
```

Next, enter the following *Mathematica* command to define the trapezoidal rule as a function of a, b, and n:

```
trap[a_,b_,n_]:=
N[(b-a)/(2n)(f[a]+Sum[2f[a+k*(b-a)/n],{k,1,n-1}]+f[b])]
```

Here **a** is the lower limit of integration in the given integral, **b** is the upper limit of integration, and **n** is the number of subdivisions of [a, b] to use. Now just enter `trap[1,3,10]`, which results in the output: 8.27635.

The estimate obtained using *Mathematica's* numerical integration command: `NIntegrate[E^Sqrt[x], {x,1,3}]` is 8.27544. This is smaller than the trapezoidal rule estimate above, by about .0007.

The command `trap[1,3,20]` yields 8.27567, which exceeds the result of *Mathematica*'s `NIntegrate` command by only about .0002.

Suppose we wish to obtain an approximation of the integral discussed above, using the trapezoidal rule, but we require our estimate be within .00001 of the actual value of the integral. The error estimate for the trapezoidal rule allows us to determine how large n should be to obtain such accuracy. In fact, the absolute value of the error in approximating $\int_a^b f(x)\,dx$ by the trapezoidal sum, is bounded above by $\dfrac{(b-a)^3 M_2}{12n^2}$, where M_2 is the maximum value of $|f''(x)|$ for $a \le x \le b$ ([1], Theorem 9.8.2). And, since M_2 and $b-a$ are fixed, the problem we face is to decide how large n must be in order to make $\dfrac{(b-a)^3 M_2}{12n^2} < .00001$.

To solve this inequality, we first use the graphing capabilities of *Mathematica* to *estimate* M_2, for we do not need the exact value of M_2, only an upper bound for that quantity.

Plot the absolute value of the second derivative over the interval [1,3]:
`Plot[Abs[f''[x]],{x,1,3}]`.

From the graph it seems clear that the largest $|f''(x)|$ can be on [1,3] is approximately 0.2 ; however, since we want to be sure not to underestimate M_2, we'll replace M_2 in our inequality by 0.22.

To solve: $|Error| < \dfrac{(b-a)^3 M_2}{12n^2} < .00001$ with $M_2 = 0.22$, enter:

```
a = 1
b = 3
M2 = .22
```

`Clear[n]`   (Type this line to make *Mathematica* forget that you previously set $n = 10$.)

`Solve[(b-a)^3 M2/(12n^2)  ==  0.00001,n]`

You should get $n = 121.106$. Since the value for n in the trapezoidal rule must be a positive integer, we conclude that if $n \geq 122$ then the trapezoidal approximation will differ from the exact value of the integral by less than .00001. Finally, check this result by comparing

`trap[1,3,122]`

and
`NIntegrate[f[x],{x,1,3}]`.

Exercises

1. Trapezoidal rule

a) Determine the number of trapezoids to use to approximate $\displaystyle\int_0^5 \frac{1}{\sqrt{1+x^5}}\,dx$ with error less than .01, using the trapezoidal rule.

b) Carry out the approximation with the number you found in part a) .

c) Compare your answer to part b) with the answer obtained by using *Mathematica's* `NIntegrate` command; assuming that value to be exact, determine the actual error of your approximation.

2. Simpson's rule

The integral $\int_0^1 e^{-x^2} dx$ cannot be calculated directly using the Fundamental Theorem of Calculus because no antiderivative of e^{-x^2} can be found. The object of this exercise is to (with the aid of *Mathematica*) use Simpson's rule to approximate this integral to within .00001 . Our first goal is the discovery of the number of terms required in the Simpson's rule approximation.

a) Define $f(x) = e^{-x^2}$ as a *Mathematica* function. (Recall that, the absolute value of the error in approximating $\int_a^b f(x) dx$ by the Simpson's rule sum is bounded above by $\dfrac{(b-a)^5 M_4}{180 n^4}$, where M_4 is the maximum value of $\left|f^{(4)}(x)\right|$ for $a \le x \le b$.)

b) Have *Mathematica* calculate $f^{(4)}(x)$ and plot the absolute value of that function on [0, 1].

c) Use the plot just obtained to find an upper bound for $M_4 = \underset{0 \le x \le 1}{\text{Max}} \left|f^{(4)}(x)\right|$.

d) Use the **Solve** command to find the smallest <u>even</u> positive integer n for which $\dfrac{M_4}{180 n^4} \le .00001$. (Remember that Simpson's rule requires the number of subintervals in the partitioning to be even.)

e) Use the value of n found in question d) to have *Mathematica* compute the Simpson's rule approximation of the integral. You can use the following *Mathematica* function for computing the Simpson's rule approximation to $\int_a^b f(x) dx$, using the partition of [a, b] into n equal subintervals:

```
Simpson[a_, b_, n_Integer]  :=
  N[(b-a)/(3n)(f[a]  +
     Sum[(3 + (-1)^(k+1))  f[a + k  (b-a)/n],{k,1,n-1}]  +
     f[b])]  /;  EvenQ[n]
```

(You needn't worry about the programming details of this command, but if you're interested, the /; EvenQ[n] at the end tells *Mathematica* to ignore this rule unless n is an even integer. Also as the counter variable k runs

from 1 to n-1 the expression $3 + (-1)^{(k+1)}$ creates the sequence
4, 2, 4, 2, . . . 4 needed as coefficients in Simpson's rule.)

f) Compare your answer to question e) with the output produced by having
Mathematica numerically integrate to evaluate the integral. The command
to enter is : `NIntegrate[f[x],{x,0,1}]`.

g) What value of n would be required in the trapezoidal rule to estimate this
same integral to within .00001 ? Compare your answer with the answer to
part d).

3. Use **NIntegrate** to attempt to evaluate $\int_0^1 \frac{1}{x^2}\sin(\frac{1}{x})dx$. Find an

antiderivative $\int \frac{1}{x^2}\sin(\frac{1}{x})dx$, and use it to explain why the **NIntegrate**

command is unable to evaluate the definite integral.

Chapter 8 L'Hopital's Rule, Improper Integrals

§1 L'Hopital's rule

The key step in finding formulas for the derivatives of the trigonometric functions was to show that $\lim\limits_{x\to 0}\dfrac{\sin(x)}{x}=1$. *Mathematica* will quickly calculate that limit in response to the command `Limit[Sin[x]/x,x->0]`. Likewise, to find $\lim\limits_{x\to\infty}\dfrac{\ln(x)}{x}$, enter: `Limit[Log[x]/x,x->Infinity]`, to which *Mathematica* will rapidly respond: `0`.

However, there are limits involving indeterminate forms which *Mathematica* will evaluate either only after an extended period of time, not at all, or incorrectly. For example, to find $\lim\limits_{x\to\infty}(1-3/x)^x$ the command `Limit[(1-3/x)^x,x->Infinity]` will, after about a minute, produce the response: e^{-3}. It takes *Mathematica* only slightly longer to correctly find $\lim\limits_{x\to\infty}(1-a/x)^x=e^{-a}$; just enter:
`Limit[(1-a/x)^x,x->Infinity]`.

Exercises

Use *Mathematica*'s `Limit` command to evaluate these limits.

1. $\lim\limits_{x\to\infty}\left(1+4/x^2\right)^x$

2. $\lim\limits_{x\to\infty}\left(\log(x)-\log(x+1)\right)$

3. (§10.3 #34) $\lim\limits_{x\to 0}\left(\dfrac{1}{x^2}-\dfrac{\cos(3x)}{x^2}\right)$

4. $\lim\limits_{x\to\pi/2}\dfrac{\sec(x)}{\tan(x)}$

Certain limits, however, are evaluated incorrectly by the `Limit` command, and others are simply returned unevaluated. To deal with such problems, we will examine the use of *Mathematica's* `Limit` command and L'Hopital's rule in combination.

Example 1 (§10.3 #55) Evaluate $\displaystyle\lim_{x\to\infty}\frac{\int_0^{2x}\sqrt{1+t^3}\,dt}{x^{5/2}}$.

Solution:

In an attempt to evaluate the limit directly, we denote the numerator in the limit by $f(x)$, and enter the command:

`f[x_]  =  Integrate[Sqrt[1+t^3],{t,0,2x}]`

Mathematica's eventual response

$$\frac{4\ x\ \text{Sqrt}[1+8x^3]}{5}+\frac{6\ x\ \text{Hypergeometric2F1}[\frac{1}{3},\frac{1}{2},\frac{4}{3},-8x^3]}{5}$$

indicates that it is able to express this antiderivative in terms of the special "hypergeometric" functions in its inventory. However, if we attempt to find the limit anyway:

`Limit[f[x]/x^(5/2),x->Infinity]`

Mathematica takes several more minutes to respond that it cannot find the limit.

To correctly find this limit, first note that both the numerator $\int_0^{2x}\sqrt{1+t^3}\,dt$ and the denominator $x^{5/2}$ approach 0 as $x\to0$. So L'Hopital's rule can be applied. The derivative of the numerator is found to be $2\sqrt{1+8x^3}$ by application of the Fundamental Theorem of Calculus and the chain rule (or by using *Mathematica*, entering: `Simplify[D[f[x],x]]`). The command

`Limit[2  Sqrt[1  +  8x^3]/(5/2  x^(3/2)),x->Infinity]`

quickly results in the correct answer: $\dfrac{2^{7/2}}{5}$, or $\dfrac{8\sqrt{2}}{5}$.

Example 2 (10.3 #67) If a denotes a positive real number then show that each of the following limits is a .

a) $\displaystyle\lim_{x\to0^+} x^{\frac{\ln a}{1+\ln x}}$ (this has the indeterminate form 0^0)

b) $\displaystyle\lim_{x\to\infty} x^{\frac{\ln a}{1+\ln x}}$ (this has the indeterminate form ∞^0)

c) $\lim\limits_{x \to 0} (x+1)^{\frac{\ln a}{x}}$ (this has the indeterminate form 1^∞)

Solution:

a) In response to the command:

`Limit[x^(Log[a]/(1+Log[x])),x->0, Direction -> -1]`
Mathematica gives the correct output: a.

b) The command:

`Limit[x^(Log[a]/(1+Log[x])),x->Infinity]`
again elicits the prompt response: a.

c) The command `Limit[(x+1)^(Log[a]/x),x->0]`
yields the output 1, which is incorrect! Thus Version 2.0 of the *Mathematica* program has a "bug" somewhere in its `Limit` command. To solve the problem correctly we can proceed as we would in solving this problem by hand: re-write $(x+1)^{\frac{\ln a}{x}}$ as $e^{\frac{\ln(x+1)\ln(a)}{x}}$. Now all the variation is in the exponent, so we look for $\lim\limits_{x \to 0} \frac{\ln(x+1)\ln(a)}{x}$. The command:

`Limit[Log[x+1](Log[a]/x),x->0]`
promptly gives the output `Log[a]`, which means (since the exponential function is continuous) that the value of our original limit is $e^{\ln(a)} = a$.

Conclusion: When using *Mathematica*'s `Limit` command to compute limits of indeterminate expressions of the form $f(x)^{g(x)}$, it is a good practice to check the result by also computing the limit of $g(x)\ln(f(x))$, as in the preceding example.

Exercises

Use L'Hopital's rule and *Mathematica* to evaluate each of the following limits.

1. $\lim\limits_{x \to 1} \dfrac{\sin(x-1)}{(x-1)^2}$

2. $\lim\limits_{x\to 0}\dfrac{\ln(\cos(ax))}{\ln(\cos(bx))}$, in terms of the constants a and b.

3. $\lim\limits_{x\to\infty}\dfrac{\int_0^x e^{t^2}\,dt}{2xe^{x^2}}$

4. $\lim\limits_{x\to\infty}\dfrac{x\int_0^x e^{t^2}\,dt}{e^{x^2}}$

5. $\lim\limits_{x\to 0}(1+x)^{\left(\frac{\ln 2}{x^2}\right)}$

§2 Improper Integrals

The *Mathematica* command
`Integrate[1/x^2,  {x,-1,2}]`
produces the result $-\dfrac{3}{2}$. This is nonsense, since the integrand $\dfrac{1}{x^2}$ is positive on the interval $[-1, 2]$. The problem is that the integrand is not continuous on the interval — it is undefined at $x = 0$ and $\lim\limits_{x\to 0}\dfrac{1}{x^2} = \infty$. The `Integrate` command applied the fundamental theorem of calculus, finding the difference between the values of an antiderivative of $\dfrac{1}{x^2}$ at the endpoints of the interval:

$-\dfrac{1}{x}\Big|_{-1}^{2} = -\dfrac{3}{2}$. But the fundamental theorem of calculus does not apply to integrals in which the integrand is unbounded on the interval of integration. To evaluate such integrals we must resort to a limiting process. Thus $\int_{-1}^{2}\dfrac{1}{x^2}\,dx$

is defined to be $\lim\limits_{s\to 0^-}\int_{-1}^{s}\dfrac{1}{x^2}\,dx + \lim\limits_{t\to 0^+}\int_{t}^{2}\dfrac{1}{x^2}\,dx$, provided both these one-sided limits

exist and are finite. If either limit fails to exist or is infinite we say $\int_{-1}^{2}\dfrac{1}{x^2}\,dx$

diverges, and in this case we do not assign a numerical value to the integral.
(If both limits are ∞ we may say $\int_{-1}^{2}\dfrac{1}{x^2}\,dx = \infty$.) The *Mathematica* command

```
Limit[Integrate[1/x^2,  {x,-1,s}],  s->0,  Direction->1]
```
produces the output `Infinity`, and also

```
Limit[Integrate[1/x^2,  {x,t,2}],  t->0,  Direction->-1]
```
yields the same output, so we conclude that $\int_{-1}^{2}\frac{1}{x^2}dx = \infty$. A plot of the graph of $y=\frac{1}{x^2}$ over the interval [–1, 2] is instructive.

```
Plot[1/x^2,  {x,-1,2},PlotRange->{0,10}]
```

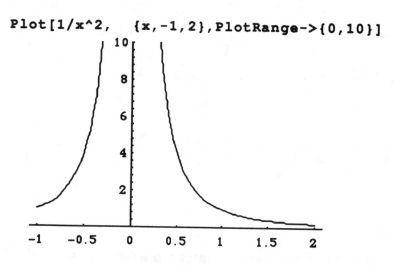

The numerical integration command `NIntegrate` can be used to approximate improper integrals, since it does not make use of the fundamental theorem of calculus. For example
```
NIntegrate[1/Sqrt[1  -  x^2],{x,-1,1}]
```
yields 3.14159, a good approximation to the exact value found by evaluating the limits: $\lim\limits_{s\to-1^+}\int_{s}^{0}\frac{1}{\sqrt{1-x^2}}dx + \lim\limits_{t\to1^-}\int_{0}^{t}\frac{1}{\sqrt{1-x^2}}dx = \frac{\pi}{2}+\frac{\pi}{2} = \pi$. The `NIntegrate` command is prepared to handle integrals like this one, in which the integrand becomes unbounded at the endpoints of the interval of integration. But if the singular point is in the interior of the interval of integration the results are not so good. For example the command
```
NIntegrate[1/x^2,  {x,-1,2}]
```
produces some messages saying that the numerical integration algorithm is converging suspiciously slowly, and then a crude numerical estimate: 9209.4.

However if we notice in advance that the integrand becomes unbounded at $x = 0$, we can modify the `NIntegrate` command to instruct *Mathematica* to integrate first from -1 to 0 and then from 0 to 2, thus making the singular point 0 an endpoint of the two intervals of integration. This can be done in a single command:

`NIntegrate[1/x^2,  {x,-1,  0,  2}]`

The resulting output $1.24066 \ 10^{3498}$ is much more in accord with the conclusion we reached by evaluating the integral using limits: $\int_{-1}^{2} \frac{1}{x^2} dx = \infty$.

Exercises

1 Show that $\int_{-1}^{2} \frac{1}{\sqrt{|x|}} dx = \int_{-1}^{2} \frac{1}{\sqrt{\sqrt{x^2}}} dx$ converges, and find its value. Check your result by using the `NIntegrate` command, using $\{x,-1, 0, 2\}$ as described above.

2. Express each of the following improper integrals using limits. Evaluate the limits and thereby determine whether the integral converges or diverges. Check your conclusions by using the `NIntegrate` command.

a) (§10.1 #22) $\displaystyle\int_{0}^{9} \frac{1}{\sqrt{9-x}} dx$

b) (§10.1 #21) $\displaystyle\int_{0}^{\pi/2} \tan x \, dx$

c) (§10.1 #1) $\displaystyle\int_{0}^{\infty} e^{-x} dx$

d) (§10.1 #17) $\displaystyle\int_{-\infty}^{\infty} \frac{x}{(x^2+3)^2} dx$

Chapter 9   Infinite Series

§1 Computing the sum of an infinite series

The following extension of the integral test allows us to compute the sum of an infinite series which has been shown to be convergent by this test.

Theorem If a positive series $\sum\limits_{k=1}^{\infty} a_k$ with decreasing terms $a_k = f(k)$ has been shown to converge by the integral test, then if the sum of the series is S, the error estimate $\int_{n+1}^{\infty} f(x)dx < S - S_n < \int_{n}^{\infty} f(x)dx$ is valid for any $n \geq 1$.

Proof:

For any $k \geq 2$ the inequality $\int_{k}^{k+1} f(x)dx \leq a_k \leq \int_{k-1}^{k} f(x)dx$ follows from the fact that the function f is decreasing on [k, k+1] and on [k-1, k], with $a_k = f(k)$. For any $n \geq 1$ adding these inequalities for k = n+1, n+2, n+3, ... gives

$$\int_{n+1}^{n+2} f(x)dx + \int_{n+2}^{n+3} f(x)dx + \int_{n+3}^{n+4} f(x)dx + ...+ \leq a_{n+1} + a_{n+2} + a_{n+3} + ...+ \leq$$

$$\int_{n}^{n+1} f(x)dx + \int_{n+1}^{n+2} f(x)dx + \int_{n+2}^{n+3} f(x)dx + ...+$$

which is equivalent to the error estimate that we wished to prove.

Example1 Compute the sum of the series $\sum\limits_{n=1}^{\infty} \dfrac{1}{n^2}$, with error less than 10^{-4}.

Solution:

Here $f(x) = \dfrac{1}{x^2}$, so $\int_{n+1}^{\infty} f(x)dx = \dfrac{1}{n+1}$ and $\int_{n}^{\infty} f(x)dx = \dfrac{1}{n}$. The theorem above then gives the inequality $S_n + \dfrac{1}{n+1} < S < S_n + \dfrac{1}{n}$. For any $n \geq 1$ this traps the sum in an interval of length $\dfrac{1}{n} - \dfrac{1}{n+1} = \dfrac{1}{n(n+1)}$. The midpoint of this interval, $M_n = S_n + \dfrac{1}{2}\left(\dfrac{1}{n+1} + \dfrac{1}{n}\right)$, must differ from S by less than $\dfrac{1}{2n(n+1)}$. So whenever n is large enough that $\dfrac{1}{2n(n+1)} < 10^{-4}$, M_n will differ from S by less than 10^{-4}. We can use *Mathematica* to perform a simple search for the smallest value of n for which $\dfrac{1}{2n(n+1)} < 10^{-4}$:

The command
```
Table[{n,   N[1/(2n(n+1))]}, {n,30,100,10}]
```
yields the output

{{30, 0.000537634}, {40, 0.000304878}, {50, 0.000196078}, {60, 0.000136612}, {70, 0.000100604}, {80, 0.0000771605}, {90, 0.0000610501}, {100, 0.000049505}}

We see that $n = 70$ is just a bit too small so we re-execute the above command, narrowing the range of values of n to $70 \leq n \leq 75$:
```
Table[{n,   N[1/(2n(n+1))]}, {n,70,75}]
```
We conclude that $\dfrac{1}{2n(n+1)} < 10^{-4}$ if $n \geq 71$. It remains only to calculate M_{71}. The

Mathematica command `Sum[N[1/n^2],{n,1,71}]` gives $S_{71} \approx 1.63095$, so

we estimate the sum of the series to be $M_{71} = S_{71} + \dfrac{1}{2}\left(\dfrac{1}{72} + \dfrac{1}{71}\right) = 1.644934979$,

and we know our error will be less than 10^{-4}. Since it is known (from other

methods) that the exact sum of the series $\displaystyle\sum_{n=1}^{\infty}\dfrac{1}{n^2}$ is $S = \dfrac{\pi^2}{6}$, our conclusion can

be checked: `N[Pi^2/6,10]` gives 1.644934067 ... , so M_{71} differs from S by less

than 10^{-6}. (In other words, in this example the value S happens to fall very

near the midpoint M_{71} of the interval $[S_{71} + \dfrac{1}{72}, S_{71} + \dfrac{1}{71}]$.

Exercises

1. Show that the series $\displaystyle\sum_{n=1}^{\infty}\dfrac{1}{n^3}$ is convergent by using the integral test. Then
use the method above to determine the sum of the series with error less than
10^{-6}. (The exact value of this series is a long-standing open question among
mathematicians.)

2. Find the sum of the series $\displaystyle\sum_{n=0}^{\infty}\dfrac{1}{1+n^2} = 1 + \dfrac{1}{2} + \dfrac{1}{5} + \dfrac{1}{10} + \dots$ with error less than 10^{-4}.

3. A simple bound for the error in truncating a series which has been shown to
be convergent by use of the alternating series test ([1], Theorem 11.7.2), is

$|S - S_n| < a_{n+1}$. Use this to find the sum of the alternating series $\displaystyle\sum_{k=1}^{\infty}(-1)^{k-1}\dfrac{k^2}{3^k}$, with

error less than 10^{-4}.

§2 Taylor Polynomials

New *Mathematica* Commands: `Series` `Normal`

In this section we'll work through an example to investigate the extent to which the Taylor polynomials of a function approximate that function.

The command `Normal[Series[f[x],{x,x0,k}]]` will give the Taylor polynomial of order k of the function $f(x)$ around a specified point x_0. For example `Normal[Series[Sin[x],{x,0,7}]]` gives the output

$$x - \frac{x^3}{6} + \frac{x^5}{120} - \frac{x^7}{5040}.$$

Exercise 1 Define the function $f(x) = \sqrt{1+x^3}$ as a *Mathematica* function. Find the Taylor polynomial of order 15 of f around the point $x_0 = 0$. Note that the only nonzero powers of x which occur are those in which the exponents are 0, 3, 6, 9, 12 and 15, i.e., multiples of 3.

We want to analyze the difference $f(x) - p_n(x) = r_n(x)$, for different values of n. To say that the sequence $\{r_n(x)\}$ converges to zero, for some fixed x-value, is equivalent to saying that for this value of x the values $p_n(x)$ approach the value $f(x)$ as n gets very large. Thus you can visualize the value $r_n(x)$ by plotting the graph of the Taylor polynomial $p_n(x)$ together with the graph of $f(x)$, over an interval centered at x_0, and examining the plot to see how far the graph of p_n differs from the graph of f at x. Here *Mathematica*'s graphical animation feature is very handy: we can make a sequence of plots, each showing the graph of $f(x)$ together with one of the curves $p_n(x)$, and then animate the sequence of images and observe how as n increases the graphs of the Taylor polynomials are related to that of f.

Example 2. Make an animation showing the graphs of the first 10 distinct Taylor polynomials of the function $f(x) = \sqrt{1+x^3}$, together with the graph of f, over the interval $[-2, 2]$.

<u>Solution:</u>

First define f as a *Mathematica* function:

```
f[x_] = Sqrt[1 + x^3]
```

Then generate the Taylor polynomials: since we saw in Exercise 1 above that only the powers of x in which the exponents are multiples of 3 occur in the Taylor polynomials of this function, the first 10 distinct Taylor polynomials of f will be $p_0, p_3, p_6, \ldots, p_{27}$. The command·

```
Do[p[x_,3k] = Normal[Series[f[x],{x,0,3k}]],{k,0,9}]
```

will define these Taylor polynomials. (The **Do** loop does not produce any output, but you can check that it has worked correctly by entering p[x,15], for example.)

We create the plots of the graphs of f and the Taylor polynomials with another **Do** loop:

```
Do[Plot[{ f[x], p[x,3k] }, {x,-2,2},
    PlotRange->{0,4},
    Epilog-> {{Text["n = ",{0.4,3}] },{Text[3k,{0.7,3}]}},
    PlotStyle->{RGBColor[1,0,0], RGBColor[0,0,1] } ],
    {k,0,9}]
```

This will produce a sequence of ten plots, with the graph of f shown in red and the graphs of the Taylor polynomials shown in blue. (Note that it takes much longer to plot the graph of $p_n(x)$ for large values of n because the number of terms in this polynomial is large, and $p_n(x)$ has to be evaluated for many values of x to plot its graph.)

After the plots have all been displayed, select the cells containing the plots and scroll up or down until one of the plots is centered on your screen. Then pull down the **Graph** menu and release the mouse button when the command **Animate Selected Graphics** is highlighted. The sequence of ten plots will be displayed in rapid succession, producing a sort of animated cartoon. (In the lower left corner of the *Mathematica* session window you'll see a row of icons which you can use to speed up or slow down the animation, show it in reverse, etc. When you want to stop the animation, just click the mouse button.)

Mathematica Note: The Epilog option in the Plot commands above allows us to include additional details in plots. In the example above we included two pieces of text: the character string "n =", starting at the point in the plot with *xy*-coordinates (0.4, 3), and the actual value of 3*k* in the plot, starting at the point (0.7, 3). (These starting points were determined after looking at a few plots to find an empty space in which this text could conveniently be placed.)

Exercise 2. After watching the animation, do you believe that the sequence $\{r_n(0.5)\}$ converges to zero, or not? For which values of *x* does it appear that $\{r_n(x)\}$ converges to 0? Explain your reasoning.

The graphical analysis above gives a good intuitive picture of the interval over which the Taylor polynomials $p_n(x)$ approximate the function *f(x)*, but much more precise information about the difference $f(x) - p_n(x)$ for any specific value of *x* can be gained by a numerical investigation. The command
`r[x_,n_]  =  f[x]  -  p[x,n]`
defines the "remainder" $r_n(x)$, and to see whether the sequence $\{r_n(x)\}$ converges to 0 for a particular value of *x*, for example *x* = 0.5, the command
`Table[N[r[0.5,   3k],10],   {k,0,9}]`
can be used. (The values *n* = 0, 3, 6, 9, 12 ... 27 are the only ones for which we have constructed `p[x,n]`.) Scanning this table we can see that the remainder appears to rapidly approach 0 when *x* = 0.5. Replacing the *x*-value by 0.9, then by 1 and finally by 1.1 gives compelling evidence that the remainder approaches 0 if $x \le 1$ but $\{r_n(x)\}$ diverges if *x* > 1. Does the sequence $\{r_n(-1)\}$ appear to converge to 0? Numerical investigation of limits of sequences sometimes gives quite convincing results, but occasionally the results can be ambiguous.

Exercise 3. Calculate the Taylor polynomials $p_n(x)$ for the function
$g(x) = \dfrac{5}{1+\frac{1}{2}x^2}$, for *n* = 0, 2, 4, ... 16. (First use the command `Clear[p]` to clear
the memory of the previously defined Taylor polynomials.) Make an animated display of the plots of the graphs of *g* and these Taylor polynomials over the interval $-3 \le x \le 3$, using the option `PlotRange->{0,   5}`. Then try selecting only the plots involving $p_0(x)$, $p_4(x)$, $p_8(x)$, $p_{12}(x)$, $p_{16}(x)$, and animating this "subsequence" of plots, which all have similar shape. (To select "disjoint" cells like this, just hold down the "command" key [marked with the ⌘ and ⌘ symbols], while clicking on the cell brackets of the desired cells.) Use the animations to estimate the interval of *x*-values for which the sequence $\{r_n(x)\}$ converges to 0. (You may want to calculate a few more Taylor polynomials $p_{20}(x)$, $p_{22}(x)$, $p_{24}(x)$, $p_{26}(x)$,..., to get a larger number of plots to animate.)

Exercise 4 Supplement your graphical analysis from Exercise 3 above by a numerical investigation of the sequence $\{r_n(x)\}$ for values of x near the endpoints of the interval over which this sequence appeared to converge to 0. Try to determine the endpoints of this interval accurate to two decimal places.

Exercise 5. (This does not involve *Mathematica*.)

Use the definition $f'(x) = \lim\limits_{h \to 0} \dfrac{f(x+h) - f(x)}{h}$ to show that for any differentiable <u>even</u> function f, the derivative f' is an <u>odd</u> function. That is, if $f(-x) = f(x)$ for all x, then show that $f'(-x) = -f'(x)$ for all x. (Similarly it can be shown that if f is an odd function then f' is even.)

It follows from Exercise 5 that if $f(x)$ is an even function, then all of its odd derivatives $f'(x)$, $f^{(3)}(x)$, $f^{(5)}(x)$, ... are odd functions. Since any odd function which is defined at $x = 0$ must have the value zero there, it follows that <u>the Taylor polynomials</u> (around the point $x_0 = 0$) <u>of an even function involve only even powers of x</u>. (The function $g(x) = \dfrac{5}{1 + \frac{1}{2}x^2}$ in Exercise 3 above is an example of such an even function; another is $f(x) = \cos(x)$.) Similarly the <u>Taylor polynomials</u> (around the point $x_0 = 0$) <u>of an odd function involve only odd powers of x</u>.

Exercise 6 As in the exercises above, some foresight is often necessary to creating the list of Taylor polynomials without duplication. For example, enter:
`Table[Normal[Series[(1-x^4)^(1/4),{x,0,n}]],  {n,1,12}].`
How could the resulting duplication be avoided? Verify your conclusion by making a list of the first 6 distinct Taylor polynomials of the function $(1 - x^4)^{\frac{1}{4}}$.

§3 Taylor Series

In this section we assume that you have studied infinite series and power series. We compute the radius of convergence of the Taylor series of a function (using the ratio or root test) and then use animations of the graphs of the function and its Taylor polynomials, as in §2, to to verify the interval of convergence.

The *Mathematica* command: `Series[f[x],{x,a,10}]` produces the power series expansion of a function f about the point $x = a$, to order at most $(x-a)^{10}$. The output of this command looks like the Taylor polynomial of f of order 10 about the point $x = a$, only it contains a final term $0[x-a]^{11}$. This final term is used to indicate all the terms in the Taylor series which involve powers of $x - a$ of degree higher than 10. The partial sums of the Taylor series, i.e., the Taylor polynomials, can be obtained by truncating the series using the `Normal` command, as described in §2 above: `Normal[Series[f[x],{x,a,10}]]`

Exercises

1. a) Define $w(x) = (1 + 4x)^{\frac{1}{3}}$ as a *Mathematica* function. What is the radius of convergence of the binomial series for $w(x)$?

   b) Use animations of the graphs of $w(x)$ and its Taylor polynomials, as in §2, to verify the interval of convergence.

2. a) Use *Mathematica* to construct the Taylor polynomial approximations of $h(x) = (x-1)^{\frac{1}{3}}$, about $a = 0$, of orders up to 10. Use a `Do` command to make a sequence of plots of the graph of $h(x)$ and its Taylor polynomials on the interval $1 \le x \le 4$, using the option `PlotRange->{-1,3}`. Animate the plots you've created. Do the Taylor polynomials approach $h(x)$ over the interval $(1,4)$, as the degree of the polynomial increases? Explain your results by finding the radius of convergence of the Taylor series of $h(x)$ about $a = 0$.

   b) Use the definition of the derivative to find $h'(1)$.

   c) Use *Mathematica* to generate Taylor polynomial approximations of $h(x)$ about $a = 1$. What happens and why ?

   d) Repeat parts a & b) of this problem with $a = 2$. Use `PlotRange->{0,2}`.

   e) Find a Taylor polynomial which approximates the function $h(x)$ over the interval $2 \le x \le 10$, with error less than .05 at every point. Use graphs and numerical computations to explain how you know your error will be small enough.

88

§4 Computations with power series

In this section we show how *Mathematica* can be used to perform computations with power series which would be tedious to carry out by hand.

Example 1 Multiply the Taylor series for e^x and for $\cos x$ to get the Taylor series expansion of the function $e^x \cos x$, up to terms of order x^{10}. Integrate the resulting series and compare it with the Taylor series for the standard antiderivative $\int e^x \cos x\, dx$ found in integral tables: $\dfrac{e^x}{2}(\cos x + \sin x)$.

Solution:

We simply compute the two series, each up to terms of order x^{10} and then multiply them:

```
expSeries  =  Series[E^x,  {x,0,10}]
cosSeries  =  Series[Cos[x],{x,0,10}]
productSeries  =  expSeries  cosSeries
```

Mathematica's **Integrate** command applies to series:

```
Integrate[productSeries,x]
```

We compare with the series expansion of $\dfrac{e^x}{2}(\cos x + \sin x)$:

```
Series[E^x/2  (Cos[x]  +  Sin[x]),{x,0,10}]
```

The results differ only in their constant terms, as required of different antiderivatives of the same function.

Mathematica note:

The output of a **Series** command is stored in the computer as a data structure called a *SeriesData* object. A *SeriesData* object of order n consists of a polynomial in x (or more generally a sum of other types of functions of x) followed by a symbol $O[x]^{n+1}$. Thus the command

```
b  =  Sum[a[k]  x^k,  {k,0,5}]  +  O[x]^6
```

creates the *SeriesData* object

$$a[0] + a[1]\ x + a[2]\ x^2 + a[3]\ x^3 + a[4]\ x^4 + a[5]\ x^5 + O[x]^6$$

To select just the polynomial part of a *SeriesData* object, we apply the *Mathematica* command **Normal**. Thus **Normal[b]** would produce the polynomial

$$a[0] + a[1] \ x + a[2] \ x^2 + a[3] \ x^3 + a[4] \ x^4 + a[5] \ x^5$$

Example 2 Divide the Taylor series for e^x by that for $\cos x$ to get the Taylor series expansion of the function $\dfrac{e^x}{\cos x}$, up to terms of order x^{10}. Integrate the result to get a series expansion of $\int \dfrac{e^x}{\cos x} \, dx$. Plot the resulting function over the interval $0 \le x \le \dfrac{\pi}{2}$.

Solution:

We follow the same procedure as Example 1, only we divide the exponential series by the Taylor series for $\cos x$.

```
quotientSeries  =  expSeries  /  cosSeries

integral  =  Integrate[quotientSeries,x]
```

The result is the first ten terms of the Taylor series for $\int \dfrac{e^x}{\cos x} \, dx$, an antiderivative which *Mathematica*'s **Integrate** command is unable to find.

Mathematica cannot plot SeriesData objects, so we apply the **Normal** command to the SeriesData object "integral", and define the resulting polynomial of degree ten as a function:

```
p[x_]  =  Normal[integral]
```
Now we can plot this function over $[0, \dfrac{\pi}{2}]$.

```
Plot[p[x],{x,0,Pi/2}]
```
Note that the function $\int \dfrac{e^x}{\cos x} \, dx$ is undefined at $x = \pi/2$, and $\lim\limits_{x \to \pi/2^-} \int \dfrac{e^x}{\cos x} \, dx = \infty$. But the Taylor polynomial p[x] we computed above is defined for all values of x. It serves as a useful approximations of $\int \dfrac{e^x}{\cos x} \, dx$ only on the interval $(-\pi/2, \pi/2)$.

Exercises

1. Use combinations of known power series to find the following limits.

   i) $\lim\limits_{x\to 0} \dfrac{e^x - 1}{x}$

   ii) $\lim\limits_{x\to 0} \dfrac{\cos x - 1}{x^2}$

   iii) $\lim\limits_{x\to 0} \dfrac{\sin x - \tan x}{(\sin x)^2}$

2. Show by squaring the power series and adding that $(\cos x)^2 + (\sin x)^2 = 1$.

3. Show that for any power series $f(x) = \sum\limits_{k=0}^{\infty} a_k x^k$, the power series expansion of $\dfrac{f(x)}{1-x}$ is $\sum\limits_{k=0}^{\infty}(a_0 + a_1 + a_2 + ... + a_k)x^k$. Use this with the familiar power series expansion of $f(x) = e^x$ to find the Taylor series expansion of $\dfrac{e^x}{1-x}$ around the origin.

4. Multiply the (binomial) power series for $\sqrt{1+x^3}$ by the Taylor series for $\cos x$, and integrate the resulting series term by term to get the terms of order less than or equal to x^{10} in the series expansion of $\int \cos x \sqrt{1+x^3}\, dx$. Plot the resulting polynomial approximation to this antiderivative over the interval $[-\pi/2, \pi/2]$. Over what interval does the polynomial provide a good approximation to the antiderivative?

5. Show that $\sum\limits_{k=1}^{\infty} \dfrac{1}{k^2} = \int_0^{\infty} \dfrac{x}{e^x - 1}\, dx = \int_0^{\infty} \dfrac{xe^{-x}}{1-e^{-x}}\, dx$ by substituting e^{-x} for u in the geometric series expansion of $\dfrac{1}{1-u}$, then multiplying by xe^{-x} and integrating term by term.

Chapter 10   Parametric Equations for Curves

§1 Polar Coordinates and Parametric Plots

The graph of a function is a curve which is useful in geometrically interpreting the properties of the function. However, many curves of geometric interest (circles, for example) are not graphs of functions of the form $y = f(x)$. A much more general means of describing plane curves is by the use of <u>parametric equations</u>: $x = x(t)$ and $y = y(t)$ where $a \le t \le b$. These specify the coordinates of a point $(x(t), y(t))$ which is thought of as tracing the curve as the auxillary variable, or <u>parameter</u> , t increases through the <u>parameter interval</u> $[a,b]$.

Mathematica has a command specifically for plotting plane curves expressed by such parametric equations: `ParametricPlot[{x[t],y[t]},{t,a,b}]` will return a plot of the curve described by the point $(x(t), y(t))$ as the parameter t increases from a to b. Note that the parameter interval $[a,b]$ is not shown in the plot -- only the path traced by the point $(x(t), y(t))$ is displayed.

The same options used in connection with the `Plot` command can be used with the `ParametricPlot` command. For example, setting `AspectRatio` to `Automatic`, yields a plot of the curve with equal scales on the two axes. In this way, a circle will look like a circle and not just an ellipse. For example, the command:
```
ParametricPlot[{2  Cos[t],2  Sin[t]},{t,0,2Pi},
     AspectRatio->Automatic]
```
results in a plot of the circle $x^2 + y^2 = 4$. In what direction is the plot swept out ?

Likewise, the command:
```
ParametricPlot[{-2  Sin[t],2  Cos[t]},{t,0,Pi},
     AspectRatio->Automatic]
```
results in a plot of the portions of the circle $x^2 + y^2 = 4$ in the second and third quadrants. In what direction is this plot swept out ?

Example 1 Obtain a plot of a portion of the curve (called a hypocycloid)

determined by the parametric equations: $\begin{cases} x = (R-r)\cos(t) + r\cos(\frac{R-r}{r}t) \\ y = (R-r)\sin(t) - r\sin(\frac{R-r}{r}t) \end{cases}$

in the case where $R = 3$ and $r = 1$. (For a geometric description see [1] §13.4 #60.)

<u>Solution:</u> Enter the following *Mathematica* commands:
```
R = 3;
r = 1;
ParametricPlot[{(R-r)Cos[t]  +  r  Cos[(R-r)t/r],
     (R-r)Sin[t]-r  Sin[(R-r)t/r]},   {t,0,Pi},
     AspectRatio->Automatic]    .
```

Try adjusting the interval over which t ranges, to obtain different portions of the curve. What is the smallest t-interval which will produce the complete curve?

Note: If you experiment with other values for R and r, you may see apparent inaccuracies in the plot when a large parameter interval is used. If this happens, try increasing the value of the PlotPoints option above its default value of 25. For example `PlotPoints->50` will cause *Mathematica* to sample 50 points in the parameter interval to start its plotting algorithm, which reduces the possibility of missing some of the details in the plot.

An important special case of the parametric representation of curves arises from a desire to plot the graph of a function $r = f(\theta)$ in polar coordinates. To do this we need only recall the equations expressing the cartesian coordinates x and y in terms of the polar coordinates: $x = r\cos(\theta)$ and $y = r\sin(\theta)$. Thus parametric equations for the polar curve are: $x = f(\theta)\cos(\theta)$ and $y = f(\theta)\sin(\theta)$. Since Greek letters cannot be used in *Mathematica* commands, we replace the symbol θ by t in our commands. Thus to instruct *Mathematica* to plot a polar graph $r = f(\theta)$, where $a \leq \theta \leq b$, just define a *Mathematica* function `f[t_]` and enter:
```
ParametricPlot[{f[t]  Cos[t],f[t]  Sin[t]},{t,a,b},
     AspectRatio-> Automatic ]  .
```

Example 2 Plot the curves
a) $r = 2\sin(\theta)$ where $0 \leq \theta \leq \pi$
b) $r = \sin(2\theta)$ where $0 \leq \theta \leq \pi/2$.
c) $r = 3(1 - \sin(\theta))$ over an appropriate interval to trace the complete curve.

<u>Solution:</u>
a) Enter the command

```
ParametricPlot[{2Sin[t]  Cos[t],2  Sin[t]^2},{t,0,Pi},
    AspectRatio -> Automatic ].
```

b) Enter the command

```
ParametricPlot[{Sin[2t]  Cos[t],Sin[2t]  Sin[t]},
    {t,0,Pi/2}, AspectRatio -> Automatic ]
```

c)   The choice of an appropriate interval over which the parameter should range to trace a polar curve exactly once, is sometimes far from clear. One cannot, for example, simply let θ vary from 0 to 2π and be confident of success. However, in this case the interval from 0 to 2π, *is* the proper choice, since $r = 3(1 - \sin\theta)$ is periodic of period 2π (plot $y = 3(1 - \sin x)$ in the cartesian plane). A larger interval would result in retracing a portion of the curve for a second time, while a smaller interval would leave part of the curve untraced. The command

```
ParametricPlot[{3(1-Sin[t])  Cos[t],3(1-Sin[t])  Sin[t]},
    {t,0,2Pi}, AspectRatio -> Automatic ]
```

will plot the complete curve.

The idea of expressing the location of a point in the plane by polar coordinates suffers very little from the fact that each point has an infinite number of representations. Though the uniqueness of representation of the xy-coordinate system is absent, many problems become more accessible when phrased in terms of polar coordinates.

Exercises

1. Plot $r = 3\sqrt{\sin(2\theta)}$ where $0 \le \theta \le \pi$, and then again where $0 \le \theta \le 2\pi$.

2. Plot each of the following curves using the given parametric representation
    a) $x(t) = 3\sin^3(t)$ and $y(t) = 3\cos^3(t)$ where $0 \le t \le 2\pi$
    b) $x(t) = \sin(2t)$ and $y(t) = 3\sin(t)$ where $0 \le t \le \pi$
    c) $x(t) = 2\cos(t) - \sin(2t)$ and $y(t) = 3\sin(t)$ where $0 \le t \le 2\pi$

3. Plot each of the following polar curves using *Mathematica's*
    ParametricPlot command (Be sure to plot the complete curve):

    a) (§13.2 #17) $r = 3 + 2\sin\theta$ b) (§13.2 #18) $r = 3 - \cos\theta$

    c) (§13.2 #21) $r = 3 + 4\cos\theta$ d) (§13.2 #25) $r^2 = -9\cos2\theta$

94

e) (§13.2 #26) $r^2 = \sin 2\theta$ f) (§13.2 #32) $r = 3\sin 2\theta$

g) (§13.2 #36) $r = \cos 5\theta$ h) (§13.2 #39) $r = 4\tan\theta$

i) (§13.2 #40) $r = 2 + 2\sec\theta$ j) (§13.2 #43) $r = 2\sin\theta\tan\theta$

Example 3 Find the area of the intersection of the regions enclosed by the curves $r = 1 + \cos(\theta)$ and $r = 1 + \sin(\theta)$.

Solution: We begin by creating a plot showing the region whose area is to be found. To do so, enter the following commands:

```
curve1=ParametricPlot[{  (1+Sin[t])Cos[t],   (1+Sin[t])Sin[t]},
    {t,0,2Pi},PlotStyle->{RGBColor[1,0,0]},
    AspectRatio-> Automatic  ]
```

```
curve2=ParametricPlot[{  (1+Cos[t])Cos[t],   (1+Cos[t])Sin[t]},
    {t,0,2Pi},PlotStyle->{RGBColor[0,0,1]},
    AspectRatio-> Automatic]
```

```
Show[curve1,curve2]
```

The region occurs in two pieces. We will find the area of the part of the region above the line $y = x$, and then (exploiting the symmetry of the region) double the result of evaluating the integral: $\frac{1}{2}\int_{\frac{\pi}{4}}^{\frac{5\pi}{4}} \left(1 + \cos(\theta)\right)^2 d\theta$. Thus the total area can be evaluated by entering:

```
Integrate[(1+Cos[t])^2,   {t,Pi/4,5Pi/4}].
```

Exercise Find the area of the region described, using *Mathematica* to make a plot showing the region, and to perform the integration.

a) (§13.3 #4) The region outside the cardioid $r = 2 - 2\cos\theta$ and inside the circle $r = 4$.

b) (§13.3 #5) The region outside the limacon $r = 2 + \sin\theta$ and inside the circle $r = 5\sin\theta$.

c) (§13.3 #6) The region enclosed by the inner loop of the limacon $r = 1 + 2\cos\theta$.

Chapter 11 Vectors

New *Mathematica* Commands: `Cross[v₁,v₂]`

§1 Vector arithmetic

Vectors are represented in *Mathematica* simply as lists of numbers. Thus the commands `v = {2, -1, 3}` and `w = {-4, 3, -1}` define two vectors in $\Re^3$. Their sum is then given by the command `v + w`, and similarly `2v - 3w` gives the desired result `{16, -11, 9}`. Also entering `v.w` produces the dot product, `-14`. By the way, note that *the dot product operation in Mathematica takes precedence over not only addition and subtraction but also over numerical multiplication, division and exponentiation.* Thus the command `v-w.w` means $\mathbf{v} - (\mathbf{w} \cdot \mathbf{w})$, not $(\mathbf{v} - \mathbf{w}) \cdot \mathbf{w}$, the expression `v.w/w.w` means $\dfrac{\mathbf{v} \cdot \mathbf{w}}{\mathbf{w} \cdot \mathbf{w}}$, and `v.w^2` means $(\mathbf{v} \cdot \mathbf{w})^2$ (When in doubt, use plenty of parentheses to make the desired order of operations clear.)

The length of a vector is expressible in terms of the dot product: $\|\mathbf{v}\| = \sqrt{\mathbf{v} \cdot \mathbf{v}}$, so the command `Sqrt[v.v]` will produce the length of the vector **v**, namely `Sqrt[14]`.

Evaluation of the cross product of two vectors requires a little more work. The designers of *Mathematica* have put it inside one of the *Mathematica* packages. To load the package, the command `Needs["LinearAlgebra`CrossProduct`"]` must be entered. You'll then have the cross product command available for the remainder of your *Mathematica* session. Continuing our example above, the command `Cross[v,w]` gives $\mathbf{v} \times \mathbf{w}$, namely `{-8, -10, 2}`. Similarly `Cross[w,v]` gives $\mathbf{w} \times \mathbf{v} = \{8, 10, -2\}$.

Example 1 Find the vector component of $\mathbf{v} = 2\mathbf{i} - \mathbf{j} + 3\mathbf{k}$ along the vector $\mathbf{w} = -4\mathbf{i} + 3\mathbf{j} - \mathbf{k}$ (sometimes called the vector projection of **v** on **w**). Find the length of this vector component (sometimes called the scalar component of **v** along **w**).

Solution:

The vector projection of **v** on **w** is given by the formula $\left(\dfrac{\mathbf{v} \cdot \mathbf{w}}{\mathbf{w} \cdot \mathbf{w}}\right)\mathbf{w}$ We compute this using the *Mathematica* command `(v.v/w.w) w`, with the output $\left\{\dfrac{28}{13}, -\dfrac{21}{13}, \dfrac{7}{13}\right\}$. The signed length of the projection of **v** on **w** is given by

$\dfrac{\mathbf{v} \cdot \mathbf{w}}{\|\mathbf{w}\|}$, which is obtained by entering the *Mathematica* command

`(v.w)/Sqrt[w.w]`. The result is $\dfrac{-14}{\text{Sqrt}[26]}$.

Example 2 Find the area of the triangle with vertices A = (2, 1, 3), B = (–1, 0 , 4), C = (3, –2, 5). Find the distance from A to the edge $\overline{BC}$.

Solution:

Define the vectors **a** = `{2, 1, 3}`, **b** = `{-1, 0, 4}` and **c** = `{3, -2, 5}` in *Mathematica*. The vector from A to B is then **b** – **a**, the vector from A to C is **c** – **a**, and the area of the triangle with corners A, B and C is half the area of the parallelogram with edges **b** – **a** and **c** – **a**. Thus the desired area is given by the formula $\dfrac{1}{2}\|(\mathbf{b}-\mathbf{a})\times(\mathbf{c}-\mathbf{a})\|$. We could compute this cross product and then find half its length, but a simpler way (especially if the computation is done by hand) is to use the formula $\|\mathbf{u}\times\mathbf{v}\|=\left\{\|\mathbf{u}\|^2\|\mathbf{v}\|^2-(\mathbf{u}\cdot\mathbf{v})^2\right\}^{\frac{1}{2}}$, which avoids computation of the cross product. Thus the area of our triangle is given by the command

`(1/2)Sqrt[(b-a).(b-a)*(c-a).(c-a)` – `(b-a).(c-a)^2]`. The output which results is $\dfrac{5\ \text{Sqrt}[6]}{2}$.

The distance from A to the edge $\overline{BC}$ is the length of the vector component of **a** – **b** orthogonal to **c** – **b**, as indicated in the sketch.

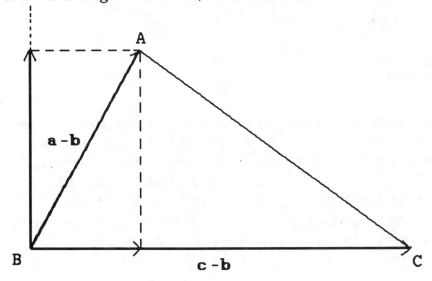

The vector component of a vector **u** orthogonal to another vector **v** is given by the formula $\mathbf{u} - \left(\dfrac{\mathbf{u}\cdot\mathbf{v}}{\mathbf{v}\cdot\mathbf{v}}\right)\mathbf{v}$, so we calculate

```
altitude  =  (a-b)  -  ((a-b).(c-b)/(c-b).(c-b))  (c-b)
```

This turns out to be $\left\{\dfrac{9}{7},\ \dfrac{13}{7},\ -\dfrac{10}{7}\right\}$. The length of this vector is then:

```
Sqrt[altitude . altitude].
```

The result is $\dfrac{5\ \mathrm{Sqrt}[2]}{\mathrm{Sqrt}[7]}$, so this is the distance from A to edge $\overline{BC}$.

Exercises

1. Let $\mathbf{u} = \{-3,\ 0,\ 1\}$, $\mathbf{v} = \{1,-1,\ 2\}$, $\mathbf{w} = \{2,\ 3,\ 0\}$. Calculate the following:
a) the angle between **u** and **v**,
b) the vector component (projection) of **u** on **w**,
c) the vector component of **u** orthogonal to **w**,
d) the length of the cross product $\mathbf{w}\times\mathbf{u}$,
e) a unit vector orthogonal to **u** and **v**,
f) the volume of the parallelopiped with edges **u**, **v** and **w**. Are the three vectors **u**, **v** and **w** coplanar?

2. Define three arbitrary vectors $\mathbf{x} = \{x_1, x_2, x_3\}$, $\mathbf{y} = \{y_1, y_2, y_3\}$ and $\mathbf{z} = \{z_1, z_2, z_3\}$, and use *Mathematica* to verify the following identities

i) $\|\mathbf{x}\times\mathbf{y}\| = \|\mathbf{x}\|^2\,\|\mathbf{y}\|^2 - (\mathbf{x}\cdot\mathbf{y})^2$

ii) $\mathbf{x}\times(\mathbf{y}\times\mathbf{z}) = (\mathbf{x}\cdot\mathbf{z})\mathbf{y} - (\mathbf{x}\cdot\mathbf{y})\mathbf{z}$

§2 Lines and planes

The parametric equations of a line in space are best understood in vector form: the line through two points P and Q is given by $\mathbf{x}(t) = \mathbf{P} + t(\mathbf{Q} - \mathbf{P})$. As t runs from 0 to 1 the point $\mathbf{x}(t)$ moves from P to Q. In fact t can be interpreted as the decimal or fractional part of the way from P to Q which $\mathbf{x}(t)$ has moved. Moreover, when t runs from 1 to 2 then $\mathbf{x}(t)$ moves further along the line, reaching the point $\|\mathbf{Q}-\mathbf{P}\|$ units beyond Q when $t = 2$. If we interpret the parameter t as time, the point $\mathbf{x}(t)$ moves from one end of the line to the other with uniform speed as t runs from $-\infty$ to ∞, reaching point P at time $t = 0$ and

Q at time $t = 1$, so its speed is $\|Q - P\|$ distance units per time unit. This can be seen clearly in an animation, as in Example 3 below.

Note that the vector equation of a line is valid regardless of the dimension of the space in which the line lies: exactly the same vector equation works for lines in $\Re^2$ and $\Re^3$. Also note that the parametric equations of a line can be read from the vector equation. Since *Mathematica* handles vectors so conveniently, we will usually use the vector form of the equation.

Example 3 Make an animation showing the point $\mathbf{x}(t)$ moving along the line $\mathbf{x}(t) = \mathbf{P} + t(\mathbf{Q} - \mathbf{P})$ as t runs from –2 to 4, using the points P = (2, –1), Q = (5, 2).
Solution:

We use the following commands

```
p = {2, -1};
q = {5, 2};
x[t_] = p + t(q - p);
a = -2;
b = 4;
frames=10;
xmin  =  Min[x[a][[1]],x[b][[1]]];
```
(*Note that* x[a] *is a vector, i.e., a list of two numbers, the first entry of which is denoted* x[a][[1]]. *Thus* xmin *is simply the smaller of the x-components of the two vectors* x[a] *and* x[b].)
```
xmax  =  Max[x[a][[1]],x[b][[1]]];
ymin  =  Min[x[a][[2]],x[b][[2]]];
ymax  =  Max[x[a][[2]],x[b][[2]]];
points  =  {PointSize[.01],RGBColor[1,0,0],Point[p],
     Point[q]};
```
This command creates a list of "graphics primitives" describing the two points p and q, both colored red.
```
Do[Show[Graphics[{points,Line[{x[a],x[t]}]}],
    PlotRange->{{xmin,xmax},{ymin,  ymax}},
    Axes->Automatic],
    {t,a,b,(b-a)/frames}]
```
The command Line[{x[a],x[t]}] *here is a graphics primitive describing the line segment from the point x[a] to x[t].*

The total range of t-values for the motion is $4 - (-2) = 6$ units, and the distance travelled is from $\mathbf{x}(-2) = \{-4, -7\}$ to $\mathbf{x}(4) = \{14, 11\}$, a distance of

99

$18\sqrt{2} \approx 25.5$. So the point $\mathbf{x}(t)$ moves $\dfrac{18\sqrt{2}}{6} = 3\sqrt{2} \approx 4.24$ distance units per

t-unit. (Equivalently, it moves $\dfrac{18\sqrt{2}}{10} \approx 2.55$ distance units per frame.)

To graph a plane, given its equation $ax + by + cz + d = 0$, the simplest method is to solve the equation for z as a function of x and y: $z = \dfrac{ax + by + d}{-c}$, provided $c \neq 0$. Then use the command

```
Plot3D[(ax + by + d)/(-c), {x, -5, 5}, {y, -5, 5},
    PlotPoints->2]
```

(The ranges of values for x and y can be changed, of course. Using the option **PlotPoints->2** tells *Mathematica* to pick only two sample points in the x and y intervals, so the resulting surface will have only $1 \times 1 = 1$ face. The default value **PlotPoints->15** would give a surface with $14 \times 14 = 196$ faces, which would require much more computing time and memory.)

If you have several planes or lines that you want to display together in a single plot, just assign each of them a name and use the **Show** command to combine the individual plots, as in Example 4 below.

Example 4 Make a plot showing the planes $2x + y - 3z = 4$ and $x - 3y + z = 7$, and calculate the acute angle between these two planes.
Solution:
The commands

```
a = Plot3D[(2x + y - 4)/3, {x, -5, 5}, {y, -5, 5},
        PlotPoints->2, DisplayFunction->Identity]

b = Plot3D[-x +3y + 7, {x, -5, 5}, {y, -5, 5},
        PlotPoints->2, DisplayFunction->Identity]
```

will create SurfaceGraphics objects representing the two planes. (The option **DisplayFunction->Identity** tells *Mathematica* to create the plot as a Graphics object, but not display it on the computer monitor.) Then

```
Show[a, b, DisplayFunction->$DisplayFunction]
```

will produce a single plot showing both planes.

To find the angle between the two planes, we use the fact that this angle is the same as the angle between appropriate normal vectors to the planes, as indicated in the following sketch.

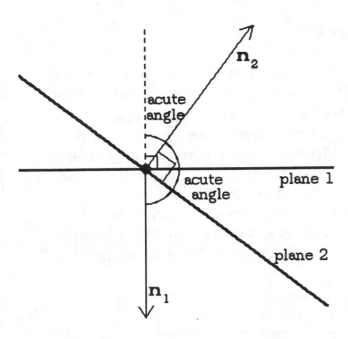

View of two planes and their normal vectors, looking down the line of intersection of the two planes -- the two acute angles are equal.

The normal vectors to the two planes can be read off from the equations of the planes, and the angle between the normal vectors

```
n1 = {2, 1, -3}
n2 = {1, -3, 1}
angle = ArcCos[n1.n2] / (Sqrt[n1.n1] Sqrt[n2.n2])]
```

The inverse cosine is left unevaluated, but we note that the dot product in the numerator, `n1.n2`, is negative, indicating that `angle` is obtuse. The acute angle is the supplementary angle, so we compute it as follows:

```
obtuseAngleInDegrees = N[angle / Degree]
acuteAngle = 180 - obtuseAngleInDegrees
```

(The constant $\frac{\pi}{180}$ is called `Degree` in *Mathematica*. To change an angle from degrees to radians, we multiply by `Degree`, and to change from radians to degrees we divide by `Degree`.)

Our calculation shows that the acute angle between the two planes is about 71.2 degrees, which agrees with the *Mathematica* plot of the two planes.

§3 Parametric equations for planes

Using the `Plot3D` command to draw planes works only on planes which are not parallel to the z-axis, i. e., planes whose equation involves the variable z. How can we plot planes such as $2x + 3y = 5$ in $\Re^3$? This is a good place to introduce the idea of vector equations (or parametric equations) for surfaces. Just as the line through a point P parallel to a vector **v** is traced out by the point $\mathbf{x}(t) = \mathbf{P} + t\mathbf{v}$ as t runs from $-\infty$ to ∞, the plane through P parallel to both **v** and a second vector **w** (**w** not parallel to **v**) is traced out by the point $\mathbf{x}(s, t) = \mathbf{P} + s\mathbf{v} + t\mathbf{w}$ as the parameters s and t independently range over all possible real values. We can picture the plane as being covered by a family of lines, one for each value of s, passing through the point $\mathbf{P} + s\mathbf{v}$ parallel to **w**, as shown in the figure.

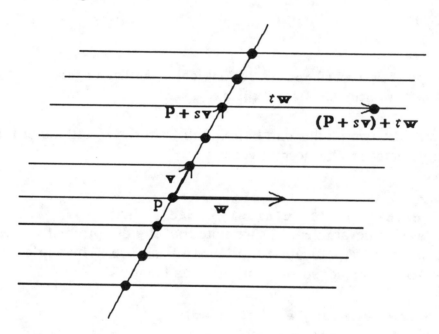

The vector equation for a plane (or any surface) $\mathbf{x}(s,t) = \{x(s,t), y(s,t), z(s,t)\}$ is equivalent to a list of three parametric equations $\begin{cases} x = x(s,t) \\ y = y(s,t) \\ z = z(s,t) \end{cases}$ for the surface.

So once we've found the vector equation we can use *Mathematica*'s `ParametricPlot3D` command to plot the surface.

Example 5 Plot the planes $2x + y - 3z = 2$ and $x - 3y = 4$ together, and find parametric equations for the line of intersection of these two planes.

Solution:

To find the vector equation of a plane we need to find a point P on the plane and two non-parallel vectors **v** and **w** which are both parallel to the plane. Perhaps the simplest way to do this is to find three non-collinear points P, Q, R on the plane. Then the vectors **v** = **Q** – **P** and **w** = **R** – **P** will satisfy our requirements. The equation of the plane has an infinite number of solutions, each corresponding to a point of the plane, so we should have no difficulty finding three of these solutions. For the plane $2x + y - 3z = 2$, for example, we could first set y and z equal to zero and solve $2x = 2$ to get a point P = (1, 0, 0), then set x and z equal to zero to get Q = (0, 2, 0) and finally set x and y equal to 0 to get R = $(0, 0, -\frac{2}{3})$. This gives the vectors **v** = **Q** – **P** = (-1, 2, 0) and **w** = **R** – **P** = $(-1, 0, -\frac{2}{3})$. Thus a vector equation for this plane is
$\mathbf{x}(s, t) = \mathbf{P} + s\mathbf{v} + t\mathbf{w} = \{1 - s - t, 2s, -\frac{2}{3}t\}$.

Finding three non-collinear points on the second plane $x - 3y = 4$ is easy too, but a bit different: for instance set $y = 0$ and $z = 0$ to get a first point (4, 0, 0), then perhaps set $y = 0$ and $z = 1$ to get a second point (4, 0, 1). To get the third point we might set $x = 1$ and $z = 0$, which yields (1, –1, 0). The resulting vector equation for this plane is $\mathbf{x}(s,t) = \{4 - 3t, -t, s\}$. Thus the following *Mathematica* commands will plot the two planes:

```
ParametricPlot3D[{{1-s-t,   2s,   -2/3 t}, {4-3t,   -t,   s}},
     {s,   -3,   3},   {t,   -3,   3},
     PlotPoints->{2,2},
     DisplayFunction  ->  Identity]
```

(The option `PlotPoints->{2,2}` tells *Mathematica* to pick only two sample points in each parameter interval, so the planes will be surfaces with only one face. When two surfaces are drawn together using the `ParametricPlot3D` command with the default value `PlotPoints->{15,15}`, each of the surfaces has 196 faces, and it takes several minutes for the plot to be displayed because so much computation is involved.)

To complete this example we must find parametric equations for the line of intersection of the two planes. The simplest way to do this is to find two points on this line, i. e., two solutions of the system of equations
$\begin{cases} 2x + y - 3z = 2 \\ x - 3y = 4 \end{cases}$. Again, there are an infinite number of solutions, corresponding to the different points on the line, so our task is easy. For example we could let

$z = 0$, which gives the system $\begin{cases} 2x + y = 2 \\ x - 3y = 4 \end{cases}$ with the unique solution $\begin{cases} x = \frac{10}{7} \\ y = -\frac{6}{7} \end{cases}$.

Next, setting $z = 1$ the equations of the two planes become $\begin{cases} 2x + y = 5 \\ x - 3y = 4 \end{cases}$, whose

solution is $\begin{cases} x = \frac{19}{7} \\ y = -\frac{3}{7} \end{cases}$. Thus the points $A = (\frac{10}{7}, -\frac{6}{7}, 0)$ and $B = (\frac{19}{7}, -\frac{3}{7}, 1)$ lie on

the line of intersection of the planes. Therefore a vector equation of this line is
$\mathbf{x}(t) = \mathbf{A} + t(\mathbf{B} - \mathbf{A}) = \{\frac{10}{7} + \frac{9}{7}t, \ -\frac{6}{7} + \frac{3}{7}t, \ t\}$. Equivalently, parametric equations
for the line are
$$\begin{cases} x = \frac{10}{7} + \frac{9}{7}t \\ y = -\frac{6}{7} + \frac{3}{7}t. \\ z = \qquad t \end{cases}$$

Check: Since the line of intersection of the two planes lies in both planes, its direction must be orthogonal to the normal vectors to each plane. Therefore it must be a multiple of the cross product of the normal vectors. The commands
`Needs["LinearAlgebra`CrossProduct`"]`
`Cross[{2, 1, -3}, {1, -3, 0}]`
yield $\{-9, -3, -7\}$. The vector $\mathbf{B} - \mathbf{A} = \{\frac{9}{7}, \frac{3}{7}, 1\}$ is indeed a multiple of the cross product, which serves as a check on our calculations.

Exercises

1. (§14.6, #19) Find parametric equations of the line through $(5, 0, -2)$, and parallel to the planes $4 - 4y + 2z = 0$ and $2x + 3y - z + 1 = 0$. Use *Mathematica* to make a plot showing the line and the two planes.

2. (§14.6, #40) Show that the line with parametric equations $\begin{cases} x = -1 + t \\ y = 3 + 2t \\ z = \quad -t \end{cases}$

    and the plane $2x - 2y - 2z + 3 = 0$ are parallel, and find the distance between them. Use *Mathematica* to make a plot showing the line and plane.

3. Show that the planes $3x - 2y + z = 6$ and $-6x + 4y - 2z = 5$ are parallel, and find the distance between them. Use *Mathematica* to plot the two planes.

Chapter 12 Vector-valued functions

§1 Limits, derivatives and integrals of vector-valued functions

Since vectors are represented simply as lists of numbers in *Mathematica*, we can define a vector-valued function of a real variable t very simply. For example, to define the function $\mathbf{f}(t) = (2+t)\mathbf{i} + t^2\mathbf{j} + \frac{1}{t}\mathbf{k}$ we just enter

`f[t_] = {2+t, t^2, 1/t}`. *Mathematica*'s commands `Limit`, `D` and `Integrate` can all be applied to vector-valued functions. Thus for the function `f` just defined the command `Limit[f[t], t->3]` produces the output $\{5, 9, \frac{1}{3}\}$, and either the command `f'[t]` or `D[f[t],t]` yields $\{1, 2t, -t^{-2}\}$. Similarly, the command

`Integrate[f[t],t]` gives the desired output $\{2t + \frac{t^2}{2}, \frac{t^3}{3}, \text{Log}[t]\}$
As was noted in Chapter 10, the graph of a function whose values are vectors in $\Re^2$ can be plotted by using the `ParametricPlot` command. The analogous command for plotting functions whose values are vectors in $\Re^3$ is `ParametricPlot3D`. For example the command
`ParametricPlot3D[f[t],{t,1,5}]`
will produce the graph of the vector-valued function $\mathbf{f}(t)$ defined above.

§2 Vector analysis of space curves and motion in space

When we view a vector-valued function as describing the position of a particle moving in $\Re^3$ (or $\Re^2$) as a function of time t, we will use the notation $\mathbf{x}(t)$, rather than $\mathbf{f}(t)$ to denote the function. (Textbook authors often use $\mathbf{r}(t)$ here, to emphasize the fact that the vector can be thought of as a radial vector out from the origin to the position of the particle.) The derivative $\mathbf{x}'(t)$ is the <u>velocity vector</u> of the particle at time t. *The length $\|\mathbf{x}'(t)\|$ gives the speed of the particle at time t, and the direction of the velocity vector, which we think of as attached at the point $x(t)$, points along the tangent line to the path of the particle at this point.* The family of velocity vectors are said to form the <u>velocity vector field</u> along the path of the particle, with one vector $\mathbf{x}'(t)$ for each instant t, attached at the corresponding point $\mathbf{x}(t)$.

The second derivative $\mathbf{x}''(t)$ describes the rate of change of the velocity at time t, so it is called the <u>acceleration vector</u> at time t, and the collection of vectors $\mathbf{x}''(t)$, each thought of as attached at the corresponding point $\mathbf{x}(t)$, forms the <u>acceleration vector field</u> along the path.

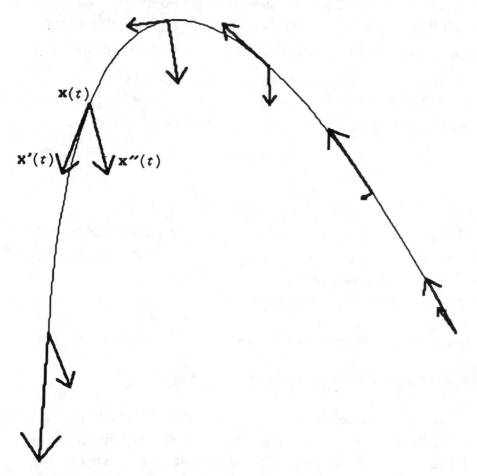

The velocity and acceleration vector fields along a path. *(The vectors $\mathbf{x}'(t)$, $\mathbf{x}''(t)$ are shown at six equally spaced times. The particle is speeding up at first, slows down when approaching the sharp bend, then speeds up again after the bend.)*

The <u>unit tangent vector</u> at $\mathbf{x}(t)$ is defined by $\mathbf{T}(t) = \dfrac{\mathbf{x}'(t)}{\|\mathbf{x}'(t)\|}$, wherever the speed $\|\mathbf{x}'(t)\|$ is nonzero. Thus the unit tangent vector field looks just like the velocity vector field, only the vectors are all adjusted to have unit length. A particle moving along the same path in such a way that its velocity vector field was $\mathbf{T}(t)$ would move with (constant) unit speed.

The <u>curvature</u> $\kappa(t)$ of the curve at $\mathbf{x}(t)$ can be computed by any one of several formulas, the simplest of which is $\kappa(t) = \dfrac{\|\mathbf{x}'(t) \times \mathbf{x}''(t)\|}{\|\mathbf{x}'(t)\|^3}$.

The <u>principal unit normal vector</u> at $\mathbf{x}(t)$ is the vector $\mathbf{n}(t) = \dfrac{\mathbf{T}'(t)}{\|\mathbf{T}'(t)\|}$, wherever $\mathbf{T}'(t) \neq 0$. We shall often refer to $\mathbf{n}(t)$ as simply the *unit normal vector* at $\mathbf{x}(t)$. Note that $\mathbf{n}(t)$, thought of as attached at the point $\mathbf{x}(t)$, always points toward the side of the path toward which the velocity vector (or equivalently the unit tangent vector $\mathbf{T}(t)$) turns, as t increases.

The acceleration vector can be decomposed into its tangential and normal components: $\mathbf{x}''(t) = \left(\|\mathbf{x}'(t)\|\right)'\mathbf{T}(t) + \left(\kappa(t)\|\mathbf{x}'(t)\|^2\right)\mathbf{n}(t)$. This formula can be made clearer by introducing a name for the speed function $\|\mathbf{x}'(t)\|$. The standard notation in the physics literature is $v(t)$, the letter v here standing for the French word for speed: *vitesse*. Suppressing the dependence of everything on t, to further simplify the notation, the formula above becomes $\mathbf{x}'' = \dfrac{dv}{dt}\mathbf{T} + \kappa v^2 \mathbf{n}$. Regardless of notation, the formula gives new insight into the motion — it says that *at any instant the tangential component of acceleration is* $\dfrac{dv}{dt}$, *the rate of change of speed of the particle, and the normal component of acceleration is the product of the curvature of the path and the* <u>square</u> *of the speed*. Doubling the speed, then, will increase the normal component of acceleration by a factor of four. Note that although the calculation of the normal vector $\mathbf{n}$ is usually quite messy, the tangential and normal components $\dfrac{dv}{dt}$ and $\kappa v^2 = \dfrac{\|\mathbf{x}'(t) \times \mathbf{x}''(t)\|}{v(t)}$ are relatively easy to compute.

Mathematica takes all the pain out of the computations involved in analyzing the motion of a particle along a curve.

Example Suppose a particle moves in the xy plane in such a way that its location at time t is $\mathbf{x}(t) = (e^{-t}\cos t)\mathbf{i} + (e^{-t}\sin t)\mathbf{j} + t\mathbf{k}$, $-1 \leq t \leq 5$.
i) Plot the path followed by the particle.
ii) Calculate the velocity, speed and acceleration of the particle as functions of t.
iii) Calculate the curvature of the path and the unit tangent and unit normal vectors at $\mathbf{x}(t)$.
iv) Verify the identity $\mathbf{x}'' = \dfrac{dv}{dt}\mathbf{T} + \kappa v^2 \mathbf{n}$, by showing that the difference between the two sides, when simplified, is the zero vector.

Solution:

First load the *Mathematica* package to compute cross products:

```
Needs["LinearAlgebra`CrossProduct`"]
```

i) Define the particle's location as a function of time, and plot it.

```
x[t_] = {E^(-t) Cos[t],E^(-t) Sin[t],t}
ParametricPlot3D[{E^(-t) Cos[t],E^(-t) Sin[t],t},{t,-1,5}]
```

ii) We define a *Mathematica* function which computes the length, or magnitude, of any vector. This will simplify our later commands.

```
magnitude[v_List] := Simplify[Sqrt[v.v],Trig->True]
```

Here, and later, we use the option `Trig->True` in the `Simplify` command, so *Mathematica* will use trigonometric identities in its attempts to simplify.

Now compute the velocity, speed and acceleration:

```
x'[t]
v[t_] = magnitude[x'[t]]
x''[t]
```

iii) This part would be a messy hand computation, but works out rather easily using *Mathematica* commands.

```
k[t_] = magnitude[Cross[x'[t],x''[t]]]/v[t]^3
T[t_] = Simplify[x'[t]/v[t],Trig->True]
```

Let's check that **T**(*t*) is a unit vector.

```
magnitude[T[t]]
n[t_] = Simplify[T'[t]/magnitude[T'[t]],Trig->True]
```

iv) Computation of the tangential and normal components of acceleration is straightforward.

```
tanComponentOfAccel[t_] = D[v[t],t]
normalComponentOfAccel[t_] = k[t] v[t]^2
```

It remains only to verify the identity $\mathbf{x}'' = \dfrac{dv}{dt}\mathbf{T} + \kappa v^2 \mathbf{n}$.

```
difference = Simplify[ x''[t] - tanComponentOfAccel[t] T[t] -
    normalComponentOfAccel[t] n[t] ]
```

This does not look like the zero vector. But note the occurrence of the expression $\mathbf{Sqrt}\left[\dfrac{1+\mathbf{E}^{2t}}{(2+\mathbf{E}^{2t})^2}\right]$ several places in the quantity $\mathbf{difference}$. This can be simplified by the command

$\mathbf{difference}$ = $\mathbf{PowerExpand[difference]}$ which tells *Mathematica* to use the rule $(a^r)^s = a^{rs}$ to simplify the expression to $\dfrac{\mathbf{Sqrt}[1+\mathbf{E}^{2t}]}{2+\mathbf{E}^{2t}}$.
This still does not give the desired result $\{0, 0, 0\}$. Several terms involve the expression and $\mathbf{Sqrt}[\mathbf{E}^{-4t}+\mathbf{E}^{-2t}]$ which *Mathematica* does not realize is equivalent to $\dfrac{\mathbf{Sqrt}[1+\mathbf{E}^{2t}]}{\mathbf{E}^{2t}}$. To replace $\mathbf{Sqrt}[\mathbf{E}^{-4t}+\mathbf{E}^{-2t}]$ by its simplified form we define a replacement rule, and then use the $\mathbf{ReplaceAll}$ command to carry out the replacement:

$\mathbf{rule} := \mathbf{Sqrt[E^{\wedge}(-4t)} + \mathbf{E^{\wedge}(-2t)} \rightarrow \mathbf{Sqrt[1 + E^{\wedge}(2t)]/E^{\wedge}(2t)\}}$
(Note the use of the "SetDelayed" equality sign := used in defining a rule. This tells *Mathematica* to just store the rule in memory. Thus no output is produced by this command. Only when $\mathbf{rule}$ is later called for will the replacement specified in the rule actually be carried out.)

$\mathbf{Simplify[\ ReplaceAll[difference\ ,\ rule]\ ]}$
Success!

Exercises

1. (§15.6, #10) If the position of a moving particle at time t is given by
     $\mathbf{x}(t) = (3t)\mathbf{i} + (2t^2)\mathbf{j} + (\ln t)\mathbf{k}$,
i) Find the velocity, speed and acceleration of the particle at times $t = 1$ and
     $t = e$. Use *Mathematica* to plot the path of the particle during the time
     interval $\frac{1}{e} \le t \le e^2$, and (by hand) draw the velocity and acceleration vectors
     at these two times on the printed plot.
ii) Use a plot of the speed as a function of time to estimate the times when the
     speed of the particle reaches its maximum and minimum values during the
     time interval $\frac{1}{e} \le t \le e^2$. Find the location of the particle at these times.

2. (§15.6, #47) Where on the path $\mathbf{x}(t) = (t^2 - 5t)\mathbf{i} + (2t + 1)\mathbf{j} + (3t^2)\mathbf{k}$ are the velocity and acceleration vectors orthogonal?

Chapter 13 Partial Derivatives

§ 1 Picturing functions of two variables

> New *Mathematica* Commands: **Plot3D**
> **ContourPlot**

Plot3D

The *Mathematica* command
Plot3D[*expr*, {x, a, b}, {y, c, d}]
makes a perspective drawing of the graph of a function or expression *expr*,
involving two variables x and y, above the specified rectangle a $\leq x \leq$ b,
c $\leq y \leq$ d in the xy-plane. Often this first plot will be satisfactory but, in case it
is not, *Mathematica* provides many options for modifying the plot. We will
discuss some of the most useful options for 3D plots in the examples below.
Further details may be found in §1.9 and §2.9 of [2].

BoxRatios

Example 1. Plot the graph of $z = x^2 - y$ over the rectangle $-1 \leq x \leq 1, 0 \leq y \leq 3$.
   Solution:
Type the command **Plot3D[x^2 - y, {x, -1, 1}, {y, 0, 3}]** to
produce the basic plot. The surface appears rather flat because the scale
chosen on the vertical z-axis is coarser than that on the x and y axes. The
interval of x-values has length 2, the y-interval has length 3, and from the
labels on the edges of the box framing the plot we see that the z-values on our
surface range from about -3 to 1, an interval of length 4. The apparent length,
width and height of the box framing the plot are specified by the **BoxRatios**
option (the 3D analogue of the **AspectRatio** option for 2D plots). The
default value **BoxRatios -> {1,1,.4}** makes the bottom of the box appear
square (i.e., the ratio of length to width is 1 to 1), and the height of the box
4/10 the length of a side of the bottom. Thus, in order to make the bottom
rectangle appear square, *Mathematica* used different scales on the x and y
axes to make the interval $-1 \leq x \leq 1$ appear to be as long as the interval
$0 \leq y \leq 3$. And the interval of z-values $-3 \leq z \leq 1$ is compressed to appear only
4/10 as long as a side of the bottom square. To see the plot using <u>equal</u> scales
on the three axes we assign the **BoxRatios** option the value {2,3,4}; that is,

the lengths of the corresponding intervals on the x, y and z axes. Enter
`Plot3D[x^2 - y, {x,-1,1}, {y,0,3}, BoxRatios->{2,3,4}]`
to see this plot. *Mathematica* provides an easier way to adjust the shape of
the box framing the plot so that the scales on the three axes will be equal: set
`BoxRatios->Automatic`. Replace the value `{2,3,4}` in the previous
command by `Automatic`, and verify that the output is unchanged. It is not
always best to use equal scales on the three axes; instead quite often it is
useful to set `BoxRatios -> {1,1,1}`, which makes the box framing the
surface a cube. Compared with the default value `{1,1,0.4}`, the plot using
`BoxRatios -> {1,1,1}` accentuates any curvature of the surface.

PlotPoints

Example 2. Plot the graph of $f(x, y) = \sqrt{1-x^2-y^2}$ over its natural domain.
Solution:
First, let's define the function f as a *Mathematica* function:
`f[x_,y_] = Sqrt[1 - x^2 - y^2].`
The natural domain of f is the disk $x^2+y^2 \le 1$, because outside this disk the
quantity $1-x^2-y^2$ is negative so its square root is undefined. The natural
domain lies inside the square $-1 \le x \le 1$, $-1 \le y \le 1$, so
`Plot3D[f[x,y], {x, -1, 1}, {y, -1, 1}]`
will plot the desired graph. When *Mathematica*'s plotting commands
encounter values of the variables outside the natural domain of the function,
these are simply ignored. However, a warning is displayed telling you that the
function values at some points are not real numbers (and therefore they
cannot be displayed in the plot). Since $z = \sqrt{1-x^2-y^2}$ is the top half of the
sphere $x^2+y^2+z^2 =1$, the graph of $f(x, y)$ should be a hemisphere. The
flattening of the graph in our plot can be eliminated by adding the option
`BoxRatios -> Automatic`, to force equal scales to be used on the three
axes, but the ragged look around the bottom of our surface is not so easily
eliminated.
   Just as a *Mathematica* plot of the graph of a function of a single variable
is really not curved, but actually is a sequence of line segments joined
together to produce the illusion of smooth curvature, the surface drawn by
Mathematica to represent the graph of a function is only an approximation to
the curved surface which constitutes the true graph. A rectangular mesh (like
a net stretched over the surface) is drawn and shaded to give the appearance
of a curved surface. To draw this mesh, *Mathematica* first inserts equally
spaced points into the intervals [-1, 1] and [-1, 1] of x and y values — the

number of these points, including the endpoints, is specified by the PlotPoints option which has the default value 15. Pairing the 15 sample x-values with each of the 15 sample y-values, we get $15 \times 15 = 225$ sample points (x, y) forming a grid on the rectangle $-1 \leq x \leq 1$, $-1 \leq y \leq 1$, and the values of the function $f(x,y)$ at these 225 sample points give the heights of the junction points of the mesh displayed on the computer screen. In our example several of the sample points fall outside the domain of f, so the corresponding portions of the mesh are not displayed, causing the jagged bottom edge of the mesh surface.

By changing the value of the PlotPoints option we can control the fineness of the rectangular mesh which approximates the true curved surface. Experiment with this by modifying the plot of the hemisphere $z = \sqrt{1-x^2-y^2}$, first using the value `PlotPoints-> 6`, and noting how crude an approximation the resulting rectangular mesh is to the hemisphere. Then repeat with `PlotPoints-> 30`. Note that the number of sample points is the <u>square</u> of the value of PlotPoints, so doubling PlotPoints from 15 to 30 increased the number of sample points by a factor of four, from 225 to 900. Now the computer has to display a mesh surface that has four times as many faces as before, which explains why it takes about four times as long to display the plot. So when you use a large value for PlotPoints, like 30 or 40, to get a high resolution plot of the graph of a function, be prepared to wait awhile!

ViewPoint

Often from the default viewpoint for the Plot3D command some interesting parts of the surface will be hidden by other parts, so you may wish to view the surface from another viewpoint. To do this, first position the cursor in your Plot3D command line where you want to insert the ViewPoint option. Then pull down the **Action** menu until the **Prepare Input** command is highlighted: a submenu will appear to the left. Carefully slide the pointer to the side into this submenu while holding the mouse button down; move the pointer down until the **3D ViewPoint Selector** command is highlighted, then release the mouse button. The 3D ViewPoint Selector dialog box will appear on the screen, with a wire-frame box in its upper left corner showing the orientation of the coordinate axes relative to the current viewpoint. Move the pointer into this box and press down the mouse button. Then when you move the mouse while holding down the button you drag the wire frame box, turning it around its center. <u>Imagine the coordinate system containing the original surface to be turning with the wire frame box</u>. When you have turned it so that from your current viewpoint (x_0, y_0, z_0) the interesting part of the surface would be visible, release the mouse button. Then simply click on the button marked **Paste ⌘A**, and the option `ViewPoint -> {x₀,y₀,z₀}` will be pasted into your Plot3D command at the point where you left your cursor. Finally, re-execute the Plot3D command (by pressing **enter**) to see the surface from the new viewpoint.

Example 3. Copy the command

`Plot3D[x^2 - y,{x, -1, 1},{y, 0, 3},BoxRatios->Automatic]`

from Example 1 above, and execute it. Suppose we want to select a viewpoint showing this slanting trough-like surface from below and to the left. To do this, first place a comma just inside the final bracket, i.e., just after the word `Automatic`. Then use the 3D ViewPoint Selector as described above to paste in a suitable value of the ViewPoint option. Execute this new command to verify that the plot of the surface from the new viewpoint is as desired. Experiment with different viewpoints until you feel confident that you could produce any desired view of the surface.

PlotRange
    Like the 2-dimensional Plot command, Plot3D selects what it thinks is an appropriate scale on the z-axis so that the most interesting parts of the surface will be shown. If this clips off parts of the surface you would like to have displayed, you can use the PlotRange option to adjust the range of z-values. For example specifying **PlotRange -> {-1,4}** will show those parts of the surface whose z-coordinates fall between -1 and 4.

AxesLabel
    The option `AxesLabel -> {"x","y","z"}` causes the axes to be labelled correctly. This option is often useful when an unusual ViewPoint has been used. *Mathematica* provides many ways to display and label the axes — we've described only the simplest labels.

Exercises.

1. Plot the graph of $g(x,y) = x^2 + y^2 + \dfrac{1}{x^2 + y^2}$ over the rectangle $-1 \le x \le 1$, $-1 \le y \le 1$. Then use the various values of the **PlotRange** option to investigate the behavior of g near the origin. You might try $\{1,100\}$, and $\{50, 100\}$, for example. Then plot the graph of g over the rectangle $-3 \le x \le 3$, $-3 \le y \le 3$. Can you now visualize the graph of g over the entire plane? Do you see which term in the formula defining $g(x,y)$ causes the spike at the origin?

2. Plot the graph of $z = xy$ over the rectangle $-3 \le x \le 3$, $-3 \le y \le 3$. Repeat, but now assign the **BoxRatios** option the value $\{1,1,1\}$. Finally, using **BoxRatios -> Automatic**, make a plot of the graph with equal scales on the three axes. If necessary, change the **ViewPoint** so that the back part of the surface will be partly visible.

3. Plot the graph of $z = e^{-(x^2+y^2)}$ over the rectangle $-3 \le x \le 3$, $-3 \le y \le 3$. Determine the range of z-values of points on the graph, and make a plot using a value for the **PlotRange** option which will show the complete graph. (That is, the top of the peak at the origin will not be clipped off.)

4. Plot the graph of $z = xye^{-(x^2+y^2)}$ over the rectangle $-3 \le x \le 3$, $-3 \le y \le 3$. Then plot this graph using a value of **PlotRange** which will show the complete graph, and using a large enough value for **PlotPoints** so that the surface will be fairly smooth. If necessary, change the **ViewPoint** and adjust the **BoxRatios** so that the interesting features of the surface are all clearly visible. Do you see how the graphs of $z = xy$ and $z = e^{-(x^2+y^2)}$ in Exercises 2 and 3 could be used to predict the shape of the graph you've found for $z = xye^{-(x^2+y^2)}$?

ContourPlot

Mathematica's **ContourPlot** command makes a shaded map of the level curves $f(x,y) = k$ of a function of two variables. (Such maps are sometimes called contour maps and the level curves are referred to as contour curves.) If f is any function or expression, **ContourPlot[f, {x, a, b}, {y, c, d}]** not only draws several level curves of f on the rectangle $a \le x \le b$, $c \le y \le d$, but also shades the regions between these curves so that low areas on the graph of f are dark gray or black and high areas on the graph are light gray or white. With such shading it is easy to distinguish between peaks and depressions on the graph by looking at a map of the level curves. Visualizing the graph from a shaded map of the level curves of a function is exactly like visualizing terrain from a topographic map.

Example 4. Use a **Plot3D** command to plot the graph of $z = x^2 + y^2$ over the rectangle $-3 \le x \le 3$, $-3 \le y \le 3$. Then make a map of the level curves of this function.

Solution:

The commands

```
Plot3D[x^2 + y^2, {x,-3,3}, {y,-3,3}, BoxRatios->{1,1,1}]
```

`ContourPlot[x^2 + y^2, {x,-3,3}, {y,-3,3}]`
will make the two plots. The circular symmetry of the graph about the z-axis is shown in the contour map by the fact that the level curves are circles centered at the origin. Notice how the shading of the two dimensional map of the level curves helps you visualize how the height varies on the three-dimensional graph.

The `ContourPlot` command, like other graphics commands, allows the user to modify the graphical output in many ways, by changing the values of various options from their default values. We will briefly discuss four of these options; others are described on pages 770-771 of [2].

ContourShading

The default value of this option is `True`, which means the contour map produced will be shaded to indicate the height on the three-dimensional graph. To suppress the shading, just use the option `ContourShading -> False`.

ContourSmoothing (or Plotpoints)

Just as the `Plot3D` command does, `ContourPlot` inserts equally spaced points into the user-given intervals of x-values and y-values. Pairing the resulting x-values and y-values gives a rectangular grid of "sample points" $\{x, y\}$, which are the foundation upon which the map of the level curves is constructed. The resulting level curves are often quite jagged, and when this happens either `ContourSmoothing -> Automatic` or `PlotPoints -> 30` will cause the number of sample points to be increased, resulting in much smoother curves. The trade-off is that smoothing the contours greatly increases the time required to display the output.

PlotRange

The `ContourPlot` command, like the `Plot3D` command, begins by determining the range of z-values of the given function at the sample points, and selects what it thinks is the most interesting interval of z-values. It then ignores all z-values outside this range. To insist that certain z-values be included, you can specify a `PlotRange` which includes these values.

Contours

The value of the option `Contours` specifies which level curves are to be drawn. `Contours->20` will produce a map with 20 level curves, corresponding to equally spaced z-values in the interval specified by `PlotRange`. The default value of `Contours` is `10`. Using a larger value gives a more detailed, but by the same token a more cluttered, map. If you prefer to specify the list of z-values for which the level curves are to be drawn, say for $z = 1, 4$, and 10, just assign `Contours -> {1,4,10}`.

Example 5 Make a more detailed map of the level curves of $z = x^2 + y^2$ around the origin. Make a separate plot showing the two level curves $x^2 + y^2 = 1$ and $x^2 + y^2 = 9$.
Solution:

To get a detailed contour map with 30 contours over the range $0 \le z \le 3$, execute the command
```
ContourPlot[x^2 + y^2, {x,-2,2}, {y,-2,2},PlotRange ->{0,3},
    Contours -> 30].
```
To make a plot showing just the two level curves $x^2 + y^2 = 1$ and $x^2 + y^2 = 9$, use the options `Contours -> {1,9}` and `ContourShading -> False` instead.

An important application of the `ContourPlot` command is to plot the graph of a single "implicitly defined function", i.e., a single level curve $f(x,y) = c$.

Example 6 Plot the graph of the curve $x^2 + 2xy^2 + 3y^3 = 4$.
Solution:

By setting `Contours -> {4}`, we force the `ContourPlot` command to plot just the single level curve we want. To get a fairly smooth curve we'll increase the `PlotPoints` value, and of course we want `ContourShading -> False`.

```
ContourPlot[x^2 + 2x y^2 + 3y^4, {x,-3,3}, {y,-3,3},
    Contours -> {4}, PlotPoints -> 30, ContourShading -> False]
```

Exercises

1. Make a map of the level curves of $f(x,y) = \cos x + \cos y$ over the rectangle $-2\pi \le x \le 2\pi$, $-2\pi \le y \le 2\pi$. Try to visualize the graph by looking at your map of the level curves of f. Check your mental image by plotting the graph, using a `Plot3D` command.

2. Make a map of the level curves of $f(x,y) = \dfrac{10x^2 y^2}{(x^2 + y^2)e^{x^2+y^2}}$ over the rectangle

$-2 \leq x \leq 2$, $-2 \leq y \leq 2$. Try to visualize the graph by looking at your map of the level curves of g. Check by plotting the graph, using a `Plot3D` command.

3. Make a map of the level curves of $g(x,y) = xye^{-(x^2+y^2)}$ over the rectangle $-3 \leq x \leq 3$, $-3 \leq y \leq 3$. Repeat, increasing the value of `Contours` to 25. What new information do you get about the graph of g? Try to visualize the graph by looking at your map of the level curves of g. Check your mental image by plotting the graph, using a `Plot3D` command.

4. The point $(0, 1)$ is said to be a "saddle point" of the function $f(x, y) = y^4 - 2y^2 - 2x^2$. Investigate the behavior of f near this point by making a plot of the level curves of f on the rectangle $-0.25 \leq x \leq 0.25$, $0.75 \leq y \leq 1.25$. Try to visualize the graph of f on this rectangle by looking at your map of the level curves. Check your mental image by plotting the graph, using a `Plot3D` command. Do you see the reason for the term "saddle point"?

5. Use a `ContourPlot` command to plot the graph of $9x^4 - 9x^2 - 4y^2 = 0$.

6. We can use a `Show` command to combine two or more *ContourGraphics* objects in a single picture. Suppose, for example, that we want a plot showing the curve $4x^2 + y^4 - 4y^2 = 0$ and the ellipse $\dfrac{x^2}{4} + y^2 = 1$ together.
The commands
```
curve1 = ContourPlot[4x^2 +  y^4  -4y^2,  {x,-2,2},  {y,-2,2},
   Contours->{0},  ContourSmoothing->Automatic,
   ContourShading -> False]
```
and

```
curve2 = ContourPlot[x^2/4  +  y^2,  {x,-2,2},  {y,-2,2},
   Contours->{1},  ContourSmoothing->Automatic,
   ContourShading -> False]
```
will plot the two curves over the rectangle $-2 \leq x \leq 2$, $-2 \leq y \leq 2$. Then

```
Show[curve1,  curve2]
```
will display these two curves, with the viewing rectangle being square, as in the individual contour plots.

Note: Parametric equations for a curve are preferable for most purposes to an "implicit function" representation, i.e., a representation as a level curve of a function of two variables. The curve $4x^2 + y^4 - 4y^2 = 0$ has parametric equations

$$\begin{cases} x = \sin 2t \\ y = 2\sin t \end{cases}, \ 0 \le t \le 2\pi, \text{ (as can be verified by substituting these expressions for } x$$

and y into the equation $4x^2 + y^4 - 4y^2 = 0$), and the ellipse $\dfrac{x^2}{4} + y^2 = 1$ can be

parametrized by $\begin{cases} x = 2\cos t \\ y = \sin t \end{cases}, \ 0 \le t \le 2\pi$. Thus the command

```
ParametricPlot[{{Sin[2t], 2Sin[t]},{2Cos[t], Sin[t]}},
        {t,0,2Pi}, AspectRatio->Automatic]
```

will make a more accurate plot of the two curves than the contour plots, and will use much less time and computer memory in doing so. Unfortunately, however, it is often difficult to find parametric equations for a given curve!

§2 Limits and Continuity

$$\lim_{(x,y) \to (x_0,y_0)} f(x,y)$$

A plot of the graph of the function f over a small rectangle centered at (x_0, y_0) can often suggest the value of the limit, or will indicate that the limit does not exist. But we must keep in mind that the mesh surface drawn by *Mathematica* is just an approximation to the true graph, using only the values of f at a grid of sample points in the rectangle. Thus conclusions drawn from such plots often need to be confirmed by analytical arguments. We will discuss two methods for investigating $\lim_{(x,y) \to (x_0,y_0)} f(x,y)$ analytically.

Method 1. Let (x, y) approach (x_0, y_0) along an arbitrary (non-vertical) line through this point: $y = y_0 + m(x - x_0)$. We simply replace y everywhere it occurs in the formula defining the function $f(x, y)$ by the expression $y_0 + m(x - x_0)$, and then take the limit of the resulting function of the single variable $x \to x_0$. If the result depends on the slope m, then the limits along different paths leading to (x_0, y_0) are not identical so $\lim_{(x,y) \to (x_0,y_0)} f(x,y)$ does not exist. However, if $\lim_{x \to x_0} f(x, y_0 + m(x - x_0))$ is independent of m it does not follow that $\lim_{(x,y) \to (x_0,y_0)} f(x,y)$ exists (see Example 5 below). Thus this method can be used to prove that a limit fails to exist, but it cannot prove the existence of the limit. Note that similar investigations can be made of the behavior of f along other curves through (x_0, y_0), to prove the nonexistence of $\lim_{(x,y) \to (x_0,y_0)} f(x,y)$. See Exercise 5 below for an example of this type.

Example 1. Investigate $\displaystyle\lim_{(x,y)\to(0,0)} \frac{x^2+y^2}{x^2-y^2}$.

Solution:

A plot of $\dfrac{x^2+y^2}{x^2-y^2}$ over a small square centered at the origin indicates that the values of this expression as (x, y) approaches the origin depend on the direction from which the approach is made. (Make such a plot!) To apply method 1, we replace y by mx and take the limit as $x\to 0$: $\displaystyle\lim_{x\to 0}\frac{x^2+m^2x^2}{x^2-m^2x^2} = \lim_{x\to 0}\frac{x^2(1+m^2)}{x^2(1-m^2)} = \frac{1+m^2}{1-m^2}$ (here the function is *constant* on each line $y = mx$). Thus as (x, y) approaches the origin along different lines, the value of $\dfrac{x^2+y^2}{x^2-y^2}$ does not approach a common value. Instead, the value depends upon the slope of the line of approach. So we conclude that $\displaystyle\lim_{(x,y)\to(0,0)} \frac{x^2+y^2}{x^2-y^2}$ does not exist.

Method 2. Examine the behavior of f on a circle of arbitrary radius r centered at (x_0,y_0). That is, replace x by $x_0+r\cos\theta$ and y by $y_0+r\sin\theta$, and find the minimum and maximum values of $f(x_0+r\cos\theta,y_0+r\sin\theta)$ for an arbitrary value of r, as θ goes from 0 to 2π. If as $r\to 0$ these minimum and maximum values of $f(x_0+r\cos\theta,y_0+r\sin\theta)$ approach a common value, then $\displaystyle\lim_{(x,y)\to(x_0,y_0)} f(x,y)$ exists and equals this value, and if not then the limit does not exist.

Example 2. Investigate $\displaystyle\lim_{(x,y)\to(0,0)} \frac{x^3}{x^2+y^2}$.

Solution:

Here $f(r\cos\theta, r\sin\theta) = \dfrac{r^3\cos^3\theta}{r^2} = r\cos^3\theta$, which ranges between $-r$ and r as θ goes from 0 to 2π. So the minimum value $-r$ and the maximum value r approach the same value 0 as $r\to 0$. Therefore this limit exists and has the value 0.

Example 3. If we apply this method to Example 1 above, we get $f(r\cos\theta,r\sin\theta)$ $= \dfrac{r^2(\cos^2\theta+\sin^2\theta)}{r^2(\cos^2\theta-\sin^2\theta)} = \dfrac{1}{\cos^2\theta-\sin^2\theta}$. Since this function of θ is independent of r (and is not constant), we conclude as before that $\displaystyle\lim_{(x,y)\to(0,0)} \frac{x^2+y^2}{x^2-y^2}$ does not exist.

Example 4. Investigate $\displaystyle\lim_{(x,y)\to(0,0)} \frac{x^3 y}{x^6 + y^2}$.

Solution:

Trying method 1 we find $f(x, mx) = \dfrac{mx^4}{x^6 + m^2 x^2} = \dfrac{mx^2}{x^4 + m^2}$, which approaches 0 as $x \to 0$. Thus if the limit exists, its value must be zero. However, the existence of the limit cannot be proven by this method. We next examine the graph of $f(x, y)$ for (x, y) near $(0, 0)$, to see if the limit appears to be zero. The command
`Plot3D[x^3 y/(x^6 + y^2), {x,-.1,.1}, {y,-.1,.1}]`
produces a surface with steep ridge near the x-axis, but which appears to level off at height zero near the origin, indicating that the limit is zero. However, when we zoom in on the origin, plotting the graph over the square $\{x, -.01, .01\}$, $\{y, -.01, .01\}$, we get essentially the same picture! That is, the ridge now comes much closer to the origin than it appeared to in the previous plot. This indicates that the behavior near the origin is too complex for the graph to be approximated accurately by *Mathematica*'s mesh surfaces. So we try method 2, examining the behavior of f on small circles centered at the origin:
$f(r\cos\theta, r\sin\theta) = \dfrac{r^4 \cos^3\theta \sin\theta}{r^6 \cos^6\theta + r^2 \sin^2\theta} = \dfrac{r^2 \cos^3\theta \sin\theta}{r^4 \cos^6\theta + \sin^2\theta}$. The maximum and minimum values of this expression for a given value of r are not obvious. But plotting
$F(r, t) = \dfrac{r^2 \cos^3 t \sin t}{r^4 \cos^6 t + \sin^2 t}$ for $0 \le t \le 2\pi$ for $r = 0.1$, $r = .01$, $r = .001$, indicates that the minimum and maximum values are -0.5 and 0.5, no matter how small r is chosen! (For example `Plot[F[.001, t], {t, 0, .00001}]` will clearly show the maximim value of 0.5, at $t \approx 10^{-6}$.) It appears that there are points arbitrarily close to the origin where the value of $f(x, y)$ is as small as -0.5 and other points where the value is as large as 0.5, which leads us to believe that the limit of f at the origin is not 0 — the limit does not exist. See Exercise 5 below for further insight into this example and a proof of the nonexistence of the limit.

Exercises: Investigate the following limits.

1. (§16.2, #15) $\displaystyle\lim_{(x,y)\to(0,0)} \frac{xy^3}{x+y}$.

2. (§16.2, #19) $\displaystyle\lim_{(x,y)\to(0,0)} \frac{x-y}{x^2 + y^2}$.

3. (§16.2, #22) $\displaystyle\lim_{(x,y)\to(0,0)} \frac{1 - \cos(x^2 + y^2)}{x^2 + y^2}$.

4. (§16.2, #27) $\displaystyle\lim_{(x,y)\to(0,0)} e^{\frac{-1}{x^2+y^2}}$.

5. (Continuation of Example 4 above) We wish to show that $\displaystyle\lim_{(x,y)\to(0,0)} \frac{x^3 y}{x^6 + y^2}$ does not exist, by showing that as (x, y) approaches $(0, 0)$ along different paths the value of $\dfrac{x^3 y}{x^6 + y^2}$ does not approach a common limit. Compute the limit as (x, y) approaches $(0, 0)$ along the curve $y = x^3$. What happens as (x, y) approaches $(0, 0)$ along $y = -x^3$? What happens as (x, y) approaches the origin along the parabola $y = x^2$? Since the limits along these paths are not all the same, we conclude that $\displaystyle\lim_{(x,y)\to(0,0)} \frac{x^3 y}{x^6 + y^2}$ does not exist.

§3 Partial Derivatives

New *Mathematica* Commands: Using the D operator for partial derivatives

The "mesh surface" *Mathematica*'s Plot3D command produces to approximate the graph of a function $f(x, y)$ is particularly useful for visualizing the partial derivatives of f. Recall that given the command Plot3D[f[x,y], {x,a,b}, {y,c,d}] the surface drawn by *Mathematica* to represent the graph of f is only an approximation to the curved surface which constitutes the true graph. A rectangular mesh (like a net stretched over the surface) is drawn and shaded to give the appearance of a curved surface. To draw this mesh, *Mathematica* first inserts equally spaced points x_i into the interval [a, b] and points y_j into [c, d] — the number of these points, including the endpoints, is specified by the PlotPoints option. The mesh lines drawn on the surface are approximations to the curves of intersection of the vertical planes $x = x_i$, and $y = y_j$ with the graph of f. Since the value of the partial derivative $f_x(x_i, y_j)$ is the slope at the point (x_i, y_j) of the curve of intersection of the plane $y = y_j$ with the graph, we can estimate this partial derivative by visually estimating the slope of this intersection curve at (x_i, y_j). Similarly, the value of $f_y(x_i, y_j)$ can be estimated by visually estimating the slope at (x_i, y_j) of the curve of intersection of the graph with the plane $x = x_i$. Also the second partial derivatives can be visually estimated by examining the mesh lines, as described in the following example.

Example 1: From a *Mathematica* plot of the graph of $f(x,y) = 2x^3y^2 + 2y + 4x$, determine the signs of $f_x(1,2)$ and $f_y(1,2)$, and determine whether the second partial derivatives $f_{xx}(1,2)$, $f_{xy}(1,2)$ and $f_{yy}(1,2)$ are positive, negative or nearly zero.

Solution:

We first define f as a *Mathematica* function:

`f[x_,y_] = 2x^3 y^2 + 2y + 4x`, then plot its graph over a small rectangle centered at the point (1, 2):

`Plot3D[f[x,y],{x,.5,1.5},  {y,1.5,2.5},  BoxRatios->{1,1,1},`
`        PlotPoints->11,  AxesLabel->{"x","y","z"}].`

We use `BoxRatios -> {1,1,1}` to accentuate the curvature of the surface, and `PlotPoints -> 11`, so that the vertical planes $x = x_i$, and $y = y_j$ will be spaced 1/10 unit apart. (To divide an interval into 10 equal subintervals requires 11 points of division, counting the endpoints, just as it takes 3 points: the two endpoints and the midpoint, to divide an interval into 2 equal subintervals.)

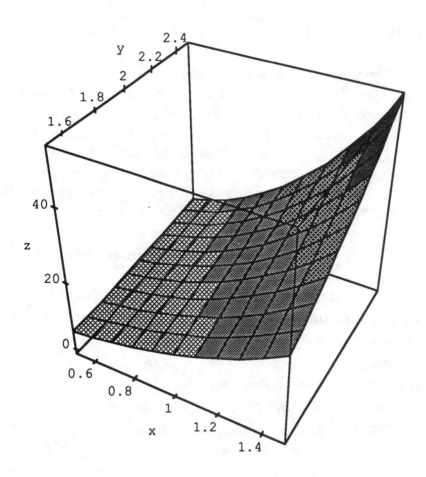

From this plot we see that the curve through the point (1, 2, 16) formed by the intersection of the plane $y = 2$ with the graph of f, rises as we move past this point along the curve in the direction of increasing x-values. So the slope $f_x(1,2)$ is positive. Moreover, it is clear that the slope increases as we move past (1, 2, 16) along this curve, i.e., the curve is concave up at this point, so $f_{xx}(1,2)$ is positive. Similarly we see from the curve of intersection of the vertical plane $x = 1$ with the surface that $f_y(1,2)$ will be positive. And since this curve appears to be almost a straight line, i.e., the slope remains nearly constant as we move past (1, 2, 16) along this curve, we conclude that $f_{yy}(1,2)$ is near 0. (Calculation shows that its exact value is 4, which is quite small on the scale used on the z-axis.) Finally, as we move past (1, 2, 16) in the direction of increasing x-values, the slope in the y-direction, $f_y(x,2)$, clearly increases, which means that $f_{xy}(1,2)$ also is positive.

Symbolic computation of partial derivatives

Mathematica's differentiation operator **D** computes partial derivatives of functions or expressions. For the function $f(x, y)$ defined above, for example, we compute $f_x(x,y)$ by the command **D[f[x,y],x]**, and the second partial derivative $f_{xx}(x,y)$ is given by either **D[f[x,y],x,x]** or **D[f[x,y],{x,2}]**. Similarly the mixed partial derivative $f_{xy}(x,y)$ is given by **D[f[x,y],x,y]**. (*Mathematica* assumes that the mixed partial derivatives in different orders are all equal.) We won't often make use of partial derivatives of higher order, but the **D** operator can compute derivatives of arbitrarily high order. To calculate $\dfrac{\partial^7 f}{\partial x^4 \partial y^3}$, for example, we would simply type **D[f[x,y],{x,4},{y,3}]**, or more clumsily, **D[f[x,y],x,x,x,x,y,y,y]**.

To evaluate a partial derivative of $f(x, y)$ at one or more points, we may define the partial derivative as a *Mathematica* function **fx[x_,y_] = D[f[x,y],x]**. Then to evaluate $f_x(1,2)$ we could simply type **fx[1,2]**. But it is not necessary to work with *Mathematica* functions when using the **D** operator. For example, the command **d1 = D[x^2 y + 3x y - 4y, x]** will produce the partial derivative of $x^2y + 3xy - 4y$ with respect to x, treating all other variables as independent of x. To evaluate such an expression at a point, say (3, –1), use the ReplaceAll command:

`ReplaceAll[d1, {x->3, y->-1}]`, or the shorter postfix `/.` equivalent: `d1 /. {x->3, y->-1}`, to get the value –9.

Exercises

1. Plot the graph of the function $z = x^2 - y^2$ over the square $0 \leq x \leq 2$, $0 \leq y \leq 2$. Use the plot to determine whether each of the partial derivatives $\dfrac{\partial z}{\partial x}$, $\dfrac{\partial z}{\partial y}$, $\dfrac{\partial^2 z}{\partial x^2}$, $\dfrac{\partial^2 z}{\partial x \partial y}$ and $\dfrac{\partial^2 z}{\partial y^2}$ are positive, negative, or nearly 0 at (1, 1). Explain your reasoning in complete sentences, as in the Example above. (Include the **AxesLabel** command, and use the ViewPoint Selector to find a suitable **ViewPoint** from which to view the surface.)

2. (§16.3, #57) Find the slope of the tangent line at (–1, 1, 5) to the curve of intersection of the surface $z = x^2 + 4y^2$ and
   a) the plane $x = -1$ b) the plane $y = 1$.

3. (§16.3, #52) Let $w = (4x - 3y + 2z)^5$. Calculate the following higher partial derivatives using *Mathematica*'s **D** operator.
   a) $\dfrac{\partial^2 w}{\partial x \partial z}$ b) $\dfrac{\partial^3 w}{\partial x \partial y \partial z}$ c) $\dfrac{\partial^4 w}{\partial z^2 \partial y \partial x}$

§ 4 Chain rules for functions of several variables

Mathematica's **D** operator correctly handles the chain rule. For example if $f(x, y)$ is an arbitrary function of two variables x and y which depend on a third variable t, say $x = x(t)$ and $y = y(t)$, then if $z = f(x, y)$, the variable z becomes a function of t as well: $z = f(x(t), y(t))$. The derivative $\dfrac{dz}{dt}$ is then given by the chain rule: $\dfrac{dz}{dt} = f_x(x(t), y(t))\, x'(t) + f_y(x(t), y(t))\, y'(t)$, which is often abbreviated

$$\frac{dz}{dt} = \frac{\partial z}{\partial x}\frac{dx}{dt} + \frac{\partial z}{\partial y}\frac{dy}{dt}.$$

This may be expressed using *Mathematica* as follows:

```
Clear[f,x,y,z,t,X,Y]
z = f[x,y];
x = X[t];
y = Y[t];
```

```
D[z,t]
```

The resulting output is

```
          (0,1)                        (1,0)
Y'[t] f        [X[t], Y[t]] + X'[t] f        [X[t], Y[t]].
```

Here $f^{(0,1)}$ denotes the partial derivative f_y of f with respect to its second variable, and $f^{(1,0)}$ denotes f_x. This works equally well in specific cases:

```
Clear[x,y,z,t]
z = x^2 y;
x = t^2 + 1;
y = 2t - 3;
D[z,t]
```
yields the output

```
                 2           2 2
4 t (-3 + 2 t) (1 + t ) + 2 (1 + t ) .
```

Similarly, if x and y are functions of two or more variables, the D operator correctly handles the chain rule:

```
Clear[f,x,y,z,s,t,X,Y]
z = f[x, y];
x = X[s,t];
y = Y[s,t];
D[z,s]
```
yields the output

```
  (0,1)                       (1,0)        (1,0)        (1,0)
f      [X[s,t],Y[s,t]] Y      [s,t] + X      [s,t] f      [X[s,t],Y[s, t]],
```

that is (rearranging the terms in standard order), $\dfrac{\partial z}{\partial s} = \dfrac{\partial z}{\partial x}\dfrac{\partial x}{\partial s} + \dfrac{\partial z}{\partial y}\dfrac{\partial y}{\partial s}$.

Example 1 Suppose $w = f(x, y, z)$, where x and y depend on z. Find an expression for $\dfrac{dw}{dz}$ in terms of partial derivatives of f and the derivatives of x and y with respect to z.

Solution:
The commands

```
Clear[f,x,y,z,w,X,Y]
w = f[x,y,z];
x = X[z];
y = Y[z];
D[w,z]  yield the output
```


(0,0,1)
f   [X[z], Y[z], z] + Y'[z] f   (0,1,0) [X[z], Y[z], z] +

          (1,0,0)
 X'[z] f   [X[z], Y[z], z]

which translates (rearranging terms in standard order) into

$$\frac{dw}{dz} = \frac{\partial w}{\partial x}\frac{\partial x}{\partial z} + \frac{\partial w}{\partial y}\frac{\partial y}{\partial z} + \frac{\partial w}{\partial z}.$$

Example 2 (§16.4,#51) let $z = f(x^2 - y^2)$. Show that $y\dfrac{\partial z}{\partial x} + x\dfrac{\partial z}{\partial y} = 0$.

Solution:

Executing the commands

```
z = f[x^2 - y^2];
y D[z,x] + x D[z,y]
```

gives 0 directly. To see in more detail what is happening, enter

`y D[z,x]` getting the output

```
            2    2
2 x y f'[x  - y ].
```

Then enter

`x D[z,y]`, to get

```
             2    2
-2 x y f'[x  - y ].
```

Adding these results gives 0.

To work the problem by hand, we would use the chain rule in the form

$$\frac{\partial z}{\partial x} = f'(x^2 - y^2)\frac{\partial}{\partial x}(x^2 - y^2) = 2x\, f'(x^2 - y^2) \text{ and}$$

$$\frac{\partial z}{\partial y} = f'(x^2 - y^2)\frac{\partial}{\partial y}(x^2 - y^2) = -2y\, f'(x^2 - y^2),$$

from which the result easily follows.

Exercises

1. (§16.4 #50) Let $z = f(xy)$. Show that $x\dfrac{\partial z}{\partial x} - y\dfrac{\partial z}{\partial y} = 0$, by evaluating the terms $x\dfrac{\partial z}{\partial x}$ and $y\dfrac{\partial z}{\partial y}$ separately. Be sure that you understand how to do this calculation by hand as well.

2. (§16.4 #56) Let $z = f(x - y, y - x)$. Show that $\dfrac{\partial z}{\partial x} + \dfrac{\partial z}{\partial y} = 0$.

3. (§16.4,#61) The equations $x = r\cos\theta$ and $y = r\sin\theta$, which relate Cartesian and polar coordinates, define r and θ implicitly as functions of x and y. Differentiate these equations with respect to x, treating r and θ as functions of x and y, and solve the resulting equations for $\dfrac{\partial r}{\partial x}$ and $\dfrac{\partial \theta}{\partial x}$, to show that

$$\frac{\partial r}{\partial x} = \cos\theta, \text{ and } \frac{\partial \theta}{\partial x} = -\frac{\sin\theta}{r}.$$

Similarly, differentiate with respect to y and show that $\dfrac{\partial r}{\partial y} = \sin\theta$ and $\dfrac{\partial \theta}{\partial y} = \dfrac{\cos\theta}{r}$.

4. (16.4 #62) Let $z = f(r, \theta)$, where r and θ are defined implicitly as functions of x and y by the equations $x = r\cos\theta$ and $y = r\sin\theta$. Use the results of Exercise 3 above to show that $\dfrac{\partial z}{\partial x} = \dfrac{\partial z}{\partial r}\cos\theta - \dfrac{1}{r}\dfrac{\partial z}{\partial \theta}\sin\theta$ and $\dfrac{\partial z}{\partial y} = \dfrac{\partial z}{\partial r}\sin\theta + \dfrac{1}{r}\dfrac{\partial z}{\partial \theta}\cos\theta$.

§5 Tangent planes, normal lines

The tangent plane to the graph of a function $f(x, y)$ at a typical point $(x_0, y_0, f(x_0,y_0))$ has the equation $z = f(x_0,y_0) + f_x(x_0,y_0)(x - x_0) + f_y(x_0,y_0)(y - y_0)$. Thus the tangent plane is also the graph of a function of x and y, so we can plot it using the Plot3D command. We can then combine the graph of f and its tangent plane in a single plot using a Show command.

Example 1. Plot the graph of $f(x, y) = \ln\sqrt{x^2 + y^2}$ and its tangent plane at (-1, 0, 0).
   Solution:
   Execute the following *Mathematica* commands:

```
f[x_,y_]  =  Log[Sqrt[x^2  +  y^2]];
surface  =  Plot3D[f[x,y],{x,-3,1},  {y,-2,2},PlotPoints->10]
fx  =  ReplaceAll[D[f[x,y],x],  {x->-1,  y->0}
fy  =  ReplaceAll[D[f[x,y],y],  {x->-1,  y->0}
```

```
tangent[x_,y_] = f[-1,0] + fx (x + 1) + fy (y - 0)
plane  =  Plot3D[tangent[x,y],{x,-3,1},  {y,-2,2},
    PlotPoints->2]
Show[surface,  plane]
```
The final **Show** command will require perhaps half a minute to produce its output, because the composite of the two surfaces is rather complex. But from an appropriate **ViewPoint**, it will be clear that the plane is tangent to the surface. (Note that **ViewPoint** is an option to the **Show** command, so if the default **ViewPoint** is not satisfactory, only the final command **Show[a, b]** must be repeated, using an appropriate **ViewPoint**.)

The normal line to the graph of f at a typical point $(x_0,\ y_0, f(x_0,y_0))$ has

parametric equations $\begin{cases} x = x_0 + f_x(x_0,y_0)t \\ y = y_0 + f_y(x_0,y_0)t \\ z = f(x_0,y_0) - t \end{cases}$. We can plot this parametric curve in

space by using the **ParametricPlot3D** command.

Example 2: Display the graph of $f(x, y) = \ln\sqrt{x^2 + y^2}$, its tangent plane at the point $(-1, 0, 0)$, and the normal line at this point.

Solution:

We simply follow the commands in Example 1 above by the commands
```
line  =  ParametricPlot3D[{-1 + fx t,  fy t,  -t},  {t,  -1,  1}]
Show[surface,  plane,  line]
```
to see the surface, tangent plane and normal line all combined in one plot.

Exercises For each surface, find an equation for the tangent plane at the given point, and find parametric equations for the normal line at this point. Use *Mathematica* to make a plot showing the surface, tangent plane and normal line together.

1. (16.5 #1) $z = 4x^3y^2 + 2y$ at $(1, -2, 12)$.

2. $z = (x^2 + 2y^2)e^{1-x^2-y^2}$ at $(\frac{1}{\sqrt{2}}, \frac{1}{\sqrt{2}}, \frac{3}{2})$.

3. (16.5 #7) $x^2 + y^2 + z^2 = 25$ at $(-3, 0, 4)$.

§6 Critical points of functions of two variables

The graph of a function of two variables, viewed either as a surface in space or as a contour map, can be used to check conclusions about the location and classification of critical points of the function reached by analytical arguments.

Example Find the critical points of the function $f(x, y) = x^4 + y^3 - 3y$, and classify each as a relative maximum, relative minimum or a saddle point.

Solution:

Computing the partial derivatives and solving the system of equations $\begin{cases} f_x(x,y) = 0 \\ f_y(x,y) = 0 \end{cases}$ in this case is so easy that we don't need *Mathematica*'s help. The critical points are (0, 1) and (0, −1). Applying the second derivative test to classify these critical points, we first compute the discriminant $D = f_{xx}(x,y) f_{yy}(x,y) - f_{xy}^2(x,y)$, which turns out to be $72x^2y$. This is zero at both critical points, so the second derivative test leads to no conclusion. However, from a contour map of f over a rectangle containing the two critical points, it is clear that (0, 1) is a local minimum and (0, −1) is a saddle point. The same conclusions can be reached by examining a suitable 3D plot of the graph of f.

Exercises For each function, find all critical points and apply the second derivative test to attempt to classify each as a relative maximum, relative minimum or a saddle point. Check your analysis by making a map of the level curves of the function in a rectangle containing all the critical points. Experiment with values of the **PlotRange** and **Contours** options to get a contour map from which you can visualize the graph of the function. Then make a plot of the graph itself, using the **Plot3D** command. Find values for the **PlotRange** and **ViewPoint** options that give a view of the surface clearly showing the character of each critical point.

1. (§16.9 #2) $f(x, y) = x^3 - 3xy - y^3$

2. (§16.9 #8) $f(x, y) = 2x^2 - 4xy + y^4 + 2$

3. (See §16.9 #23) $f(x, y) = 3xe^y - x^3 - e^{3y}$. Show analytically that f has only one critical point, a relative maximum, but it has no absolute maximum. Try to show in a *Mathematica* plot that the graph has a relative maximum but no saddle point.

129

4. (See §16.9 #24) $f(x, y) = 4x^2 e^y - 2x^4 - e^{4y}$. Show analytically that f has only two critical points, both relative maxima. Try to show in a *Mathematica* plot that the graph has two relative maxima, but no saddle point between them.

§ 7 Lagrange Multipliers

The most difficult part of solving a constrained optimization problem using the method of Lagrange multipliers is usually the algebra involved in solving the systems of equations which arise. These equations are often highly nonlinear in the basic variables x, y, z, perhaps even involving transcendental functions like logarithms, exponentials and the trigonometric functions. Thus *Mathematica*'s `Solve` command may not be able to solve these equations. In such cases one must resort to the `FindRoot` command to numerically approximate solutions of the equations. However, most textbook problems involve only polynomial functions, so the `Solve` command will find all solutions.

Example. Find the coordinates of the highest point on the curve of intersection of the plane $2x - 3y + 4z = 12$ and the sphere $x^2 + y^2 + z^2 = 25$.
Solution:

Since the z-coordinate on the plane is related to the x and y coordinates by the equation $z = \dfrac{12 - 2x + 3y}{4}$, the problem is to find the maximum value of the

function $f(x, y) = \dfrac{12 - 2x + 3y}{4}$ subject to the constraint $x^2 + y^2 + \left(\dfrac{12 - 2x + 3y}{4}\right)^2 = 25$.

Taking gradients yields the two equations
$$\begin{cases} -\dfrac{2}{4} = \lambda\left(2x - \dfrac{12 - 2x + 3y}{4}\right) \\ \dfrac{3}{4} = \lambda\left(2y + \dfrac{3(12 - 2x + 3y)}{8}\right) \end{cases}.$$ We must

solve the system of three equations in x, y and λ consisting of the constraint equation together with these two equations. The following commands will do this:

```
equations  =  {-2/4  ==  L  (2x  -  (12-2x+3y)/4),
                 3/4  ==  L  (2y  +  3(12-2x+3y)/8),
                 x^2  +  y^2  +  ((12-2x+3y)/4)^2  ==  25};

Solve[equations,  {x,y,L}]
```

N[%]

(The exact solutions produced by the **Solve** command involve messy square roots and fractions, so we apply the **N** command to get numerical approximations.)

The resulting output is

```
{{L -> -0.0747917, y -> -4.00768, x -> 2.67179},
 {L -> 0.0747917, y -> 1.52492, x -> -1.01662}},
```

from which we conclude that the point we are looking for is either (2.67179, –4.00768) or (–1.01662, 1.52492). Evaluating f at these two points we find that the first gives the minimum of f and the second gives the maximum, about 4.65200. So the highest point on the curve of intersection is approximately (–1.01662, 1.52492, 4.65200).

Exercises For problems 1, 2 and 3, work out the Lagrange multiplier equations; then use *Mathematica*'s **Solve** command to solve the system consisting of these equations together with the constraint equation. Finally, determine which solutions give the maximum value of $f(x, y)$ and which give the minimum value, subject to the constraint.

1. (§16.10 #4) $f(x, y) = x - 3y - 1$; constraint: $x^2 + 3y^2 = 16$.

2. (§16.10 #6) $f(x, y) = 3x + 6y + 2z$; constraint: $2x^2 + 4y^2 + z^2 = 70$.

3. (§16.10 #12) Find the points on the surface $xy - z^2 = 1$ that are closest to the origin.

4. Find the points on the sphere $x^2 + y^2 + z^2 = 9$ at which the value of the expression $x^2 y$ is largest.

Chapter 14 Multiple Integrals

§ 1 Double integrals

> New *Mathematica* commands: Using $\begin{array}{c}\texttt{Integrate}\\ \texttt{NIntegrate}\end{array}$ to evaluate double integrals

The plot of the graph of a function $f(x, y)$ over a rectangle R with sides parallel to the x and y axes can be used to visualize the double integral $\iint\limits_{R} f(x,y)dA$ in much the same way that we use the graph of a function of a single variable $g(x)$ over an interval [a,b] to visualize the integral $\int_a^b g(x)dx$. The difference is that, if f is non-negative on R, the double integral represents the *volume* of the solid region between R and the graph of f, while $\int_a^b g(x)dx$ gives the *area* of the region between the interval [a, b] and the graph of g. As with single integrals, if the integrand f is negative on part of R then $\iint\limits_{R} f(x,y)dA$ gives a difference: the volume of the solid lying between the graph and the part of R where f is positive, minus the volume of the solid lying between the graph and the part of R where f is negative. By using the plot produced by *Mathematica*'s **Plot3D** command, we can visually estimate these volumes and thus get a crude approximation of the value of the integral. Such a visual approximation can serve as a useful check on an analytical calculation of the integral.

Example 1: Estimate $\int_0^3\int_0^4 \sqrt{25-x^2-y^2}\,dy\,dx$ graphically.

Solution:

This iterated integral equals the double integral $\iint\limits_{R}\sqrt{25-x^2-y^2}$ where R is the rectangle $0 \le x \le 3, 0 \le y \le 4$. So we plot the graph of $\sqrt{25-x^2-y^2}$ over R:
Plot3D[Sqrt[25 - x^2 - y^2],{x,0,3}, {y,0,4},
   BoxRatios->Automatic].
From the plot it appears that the average height of the graph is about 4, so the volume under the graph is about 4 times the area of R, i.e., about 48. Since the function $\sqrt{25-x^2-y^2}$ is positive throughout R, this volume is the value of the double integral.

Mathematica's `Integrate` command can be used to evaluate iterated integrals, provided that the program is able to find the necessary antiderivatives.

Example 2. Evaluate $\int_0^3 \int_{-1}^4 2x^2 y - xy^2 \, dx \, dy$.

Solution:
Just execute the command
`Integrate[2x^2 y - x y^2, {x,-1,4},{y,0,3}]`
to get the exact result $\frac{255}{2}$.

Sometimes *Mathematica* is unable to evaluate the necessary antiderivatives, but applying the "numerically evaluate" command `N` will still lead to a decimal approximation.

Example 3. Evaluate the integral in Example 1.

Solution:
The command
`Integrate[Sqrt[25 - x^2 - y^2],{x,0,3}, {y,0,4}]`
will, after about two minutes, return the output

$$\frac{9\,Pi}{2} + Integrate[\frac{25\,ArcSin[\frac{4}{(25-x^2)^{1/2}}]}{2} - \frac{x^2\,ArcSin[\frac{4}{(25-x^2)^{1/2}}]}{2}, \{x,\ 0,\ 3\}]$$

This means that the program was able to perform the first antidifferentiation with respect to y, $\int_0^4 \sqrt{25 - x^2 - y^2} \, dy$, but it was unable to antidifferentiate the resulting function of x. However, if we simply type `N[%]`, then after about a minute the numerical approximation 48.1711 is produced. Since we know from Example 1 that the value should be about 48, we can be confident that *Mathematica*'s result is correct.

Note: We can also evaluate iterated integrals numerically, using the `NIntegrate` command with the option `Method -> MultiDimensional`. For example the command

`NIntegrate[Sqrt[25 - x^2 - y^2],{x,0,3}, {y,0,4},`
`    Method -> MultiDimensional]`
will approximate the integral in Example 1 without any antidifferentiation.

133

As with single integrals, it is possible to find examples where *Mathematica*'s `Integrate` command yields incorrect results. Likewise, the numerical integration command `NIntegrate` can make mistakes. Thus a graphical check such as we performed in Example 1 is a good precaution.

Mathematica's `Integrate` command can handle iterated integrals in which the endpoints of integration for the inner integral depend on the outer integration variable. Thus it can be used to evaluate double integrals over nonrectangular regions.

Example 4. Evaluate $\iint\limits_{R} x^2 y \, dA$, where R is the triangle bounded by the x and y axes and the line $x + y = 4$

Solution:

The command `Plot[4-x,{x,0,4}]` gives a picture of the triangle:

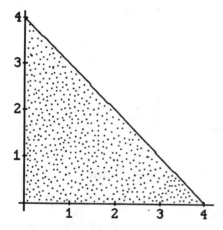

Figure 1 Graph of $0 \le y \le 4 - x, \quad 0 \le x \le 4$

The line intersects the axes at (4, 0) and (0, 4), so
$$\iint\limits_{R} x^2 y \, dA = \int_0^4 \int_0^{4-x} x^2 y \, dy \, dx.$$ The *Mathematica* command

`Integrate[x^2 y, {x,0,4}, {y,0,4-x}]` will produce the result $\dfrac{256}{15}$.

Note that <u>the order in which the variable ranges are specified is the same as on the integral signs</u>: first the outer limits {x, 0, 4}, then the inner limits {y, 0, 4 - x}.

Exercises

1. (§17.1 #28) Plot the graph of the plane $z = 2x + y$ over the rectangle $3 \le x \le 5$, $1 \le y \le 2$. Express the volume of the solid region above this rectangle and below this plane as a double integral. Then evaluate the integral.

2. (§17.1 #30) Let G be the solid region in the first octant bounded by the coordinate planes and the three planes $z - y = 0$ and $z = -2y + 6$ and $x = 5$. Plot the first two planes, using Plot3D commands, and combine the two plots by using a Show command. (First, what range of values for x and y would be appropriate here?) When you have a good plot, express the volume of G as the sum or difference of two double integrals, and evaluate these integrals.

3. (§17.2 #6) Sketch the region R such that the iterated integral $\int_{-1}^{1} \int_{-x^2}^{x^2} x^2 - y\, dy\, dx$ equals the double integral $\iint_R x^2 - y\, dA$. Evaluate the iterated integral by hand, and then check your answer using *Mathematica*'s Integrate command.

4. (§17.2 #19) Evaluate the double integral $\iint_R x(1+y^2)^{-\frac{1}{2}}\, dA$, where R is the region in the first quadrant enclosed by the curves $y = x^2$, $y = 4$, and $x = 0$. (Hint: Choose your order of integration carefully.)

§ 2 Double integrals in Polar Coordinates

Recall the general formula for changing to polar coordinates in a double integral: $\iint_{R_{xy}} f(x,y)\, dA = \iint_{R_{\theta r}} f(r\cos\theta, r\sin\theta)\,|r|\, dA$, where $R_{\theta r}$ is the region in the θr-plane which corresponds to the region R_{xy}. That is, as (θ, r) varies over the region $R_{\theta r}$, the point $(r\cos\theta, r\sin\theta)$ covers the region R_{xy}, and distinct points (θ, r) in $R_{\theta r}$ correspond to distinct points $(r\cos\theta, r\sin\theta)$ in R_{xy}.

There are two reasons one might change to polar coordinates — either because the region $R_{\theta r}$ might have simpler geometry than R_{xy} or because the integrand $f(r\cos\theta, r\sin\theta)$ might be easier to antidifferentiate than $f(x, y)$.

135

Example Evaluate $\iint\limits_{R_{xy}} \dfrac{4}{\sqrt{16-x^2-y^2}}\, dA$, where R_{xy} is the lower half of the disk

$(x-2)^2 + y^2 = 4$.

Solution:

Here both the region of integration and the integrand simplify when we change to polar coordinates. The integrand becomes $\iint\limits_{R_{\theta r}} \dfrac{4|r|}{\sqrt{16-r^2}}\, dA$. Expanding the equation $(x-2)^2 + y^2 = 4$ and changing to polar coordinates gives $x^2 - 4x + y^2 = 0$, $r^2 - 4r\cos\theta = 0$, or $r(r-4\cos\theta) = 0$. Thus the region $R_{xy} = \left\{(x,y)\,\middle|\,(x-2)^2 + y^2 \le 4 \text{ and } y \le 0\right\}$ corresponds to the region $R_{\theta r} = \left\{(r,\theta)\,\middle|\,r(r-4\cos\theta) \le 0 \text{ and } r\sin\theta \le 0\right\}$. We can picture the region $R_{\theta r}$ using a *Mathematica* Plot command:

```
Plot[4Cos[t],  {t,  -Pi/2,  3Pi/2},
    Ticks->{{-Pi/2,0,Pi/2,Pi,3Pi/2},Automatic}]
```
(We've added stippling and some labels to the *Mathematica* output (using the MacPaint drawing program) for later reference.)

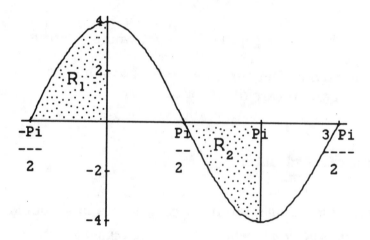

Figure 2 Cartesian graph of $r = 4\cos\theta$

Note: It is important to distinguish between the *cartesian* graph of the equation $r = 4\cos\theta$, which is produced by *Mathematica*'s Plot command, and the *polar* graph of this equation, which is produced by the *Mathematica* command
```
ParametricPlot[{(4Cos[t]) Cos[t],  (4Cos[t]) Sin[t]},
    {t,-Pi/2,  Pi/2}].
```
The cartesian graph is a boundary curve of $R_{\theta r}$, while the polar graph bounds the region R_{xy}.

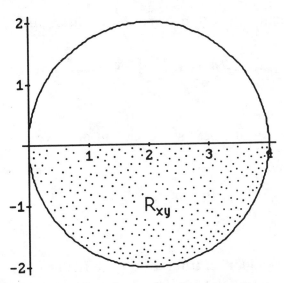

Figure 3 Polar graph of $r = 4\cos\theta$

If $r = 4\cos\theta$, as the polar angle θ runs from $-\pi/2$ to $\pi/2$, the point $(x, y) = (r\cos\theta, r\sin\theta)$ goes once around the circle $(x-2)^2 + y^2 = 4$. The lower half of the circle is traced out as θ runs from $-\pi/2$ to 0, because over this range $y = r\sin\theta$ is negative. Also as θ runs from $\pi/2$ to $3\pi/2$, if the point $(x, y) = (r\cos\theta, r\sin\theta)$ goes again around the circle, this time with the lower half of the circle being traced out as θ runs from $\pi/2$ to π. Thus either the region marked R_1 in Figure 2 or that marked R_2 can serve as the region $R_{\theta r}$ in the formula for changing to polar coordinates in a double integral. On R_1, r is positive so

$$\iint_{R_{xy}} \frac{4}{\sqrt{16-x^2-y^2}}\, dA = \iint_{R_1} \frac{4r}{\sqrt{16-r^2}}\, dA = \int_{-\pi/2}^{0}\int_{0}^{4\cos\theta} \frac{4r}{\sqrt{16-r^2}}\, dr\, d\theta.$$

Mathematica's `Integrate` command incorrectly evaluates this integral, so we will work it out by hand.

First, $\displaystyle\int_{0}^{4\cos\theta} \frac{4r}{\sqrt{16-r^2}}\, dr = -4\sqrt{16-r^2}\,\Big|_{0}^{4\cos\theta} = -4\sqrt{16\sin^2\theta} + 16 = -16|\sin\theta| + 16.$

Since $\sin\theta$ is negative for $-\pi/2 \le \theta \le 0$, then $-16|\sin\theta| = 16\sin\theta$. Thus

$$\int_{-\pi/2}^{0}\int_{0}^{4\cos\theta} \frac{4r}{\sqrt{16-r^2}}\, dr\, d\theta = 16\int_{-\pi/2}^{0}\sin\theta + 1\, d\theta = -16 + 8\pi.$$

(The `Integrate` command correctly evaluates the integral with respect to r:
`Integrate[4r/Sqrt[16 - r^2], {r,0, 4Cos[t]}]`
yields the output
`16 - 16 Sqrt[Sin[t]^2]`, which is equivalent to our $-16|\sin\theta| + 16$ above. But the second integration from `-Pi/2` to `0` yields the incorrect value `16 + 8 Pi`.)

We would get the same result if we used the region R_2 as $R_{\theta r}$ in changing to polar coordinates. On R_2, r is negative so

$$\iint_{R_{xy}} \frac{4}{\sqrt{16-x^2-y^2}}\, dA = \iint_{R_2} \frac{-4r}{\sqrt{16-r^2}}\, dA = \int_{\pi/2}^{\pi} \int_{4\cos\theta}^{0} \frac{-4r}{\sqrt{16-r^2}}\, dr\, d\theta = \int_{\pi/2}^{\pi} \int_{0}^{4\cos\theta} \frac{4r}{\sqrt{16-r^2}}\, dr\, d\theta.$$

Now

$$\int_{0}^{4\cos\theta} \frac{4r}{\sqrt{16-r^2}}\, dr = -16|\sin\theta| + 16, \text{ as above, but now since } \sin\theta \text{ is positive}$$

for $\pi/2 \leq \theta \leq \pi$, then $-16|\sin\theta| = -16\sin\theta$. Thus

$$\int_{\pi/2}^{\pi} \int_{0}^{4\cos\theta} \frac{4r}{\sqrt{16-r^2}}\, dr\, d\theta = 16\int_{\pi/2}^{\pi} -\sin\theta + 1\, d\theta = -16 + 8\pi.$$

Exercises For Exercises 1 & 2, sketch the region of integration R_{xy} using *Mathematica*. Change to polar coordinates and sketch the corresponding region $R_{\theta r}$. Express the double integral as an iterated integral in polar coordinates, and evaluate it.

1. (§17.3 #20) $\displaystyle\iint_{R_{xy}} \sqrt{9-x^2-y^2}\, dA$, where R_{xy} is the region in the first quadrant within the circle $x^2 + y^2 = 9$.

2. (§17.3 #22) $\displaystyle\iint_{R_{xy}} 2y\, dA$, where R_{xy} is the region in the first quadrant bounded above by the circle $(x-1)^2 + y^2 = 1$ and below by the line $y = x$.

3. Let G denote the solid region bounded on the sides by the cylinder $(x-1)^2 + y^2 = 1$, above by the paraboloid $z = 16 - x^2 - y^2$, and below by the plane $z = 0$. Express the volume of G as a double integral over a region R_{xy}. Change to polar coordinates and express the volume of G as a double integral over a region $R_{\theta r}$. Evaluate the volume of G.

§3 Parametric equations for surfaces

We've used double integrals to find volumes of certain solids, and in the next section we will be integrating functions over solid regions. To determine the correct iterated integrals to evaluate for this purpose one must be able to visualize the solids. We typically visualize solids by means of their boundary surfaces, so we will need a flexible method for picturing surfaces in space. *Mathematica*'s `ParametricPlot3D` command gives us a powerful tool for this purpose, but to make use of this command to plot a surface, one must first find parametric equations for the surface.

Recall that a curve is most naturally described by specifying the coordinates of a typical point (x, y, z) on the curve in terms of a single auxiliary variable, or parameter: $\begin{cases} x = x(t) \\ y = y(t), \\ z = z(t) \end{cases}$ $a \le t \le b$. As the parameter t runs through the parameter interval $[a, b]$, the point $(x(t), y(t), z(t))$ moves along the curve. For example, the line segment from a point $P = (x_0, y_0, z_0)$ to another point $Q = (x_1, y_1, z_1)$ is, in vector form, $\mathbf{x} = \mathbf{P} + t(\mathbf{Q} - \mathbf{P})$, $0 \le t \le 1$, or in standard parametric form $\begin{cases} x = x_0 + t(x_1 - x_0) \\ y = y_0 + t(y_1 - y_0), \\ z = z_0 + t(z_1 - z_0) \end{cases}$ $0 \le t \le 1$. Thus the parameter t in this example can be interpreted as the fraction (or decimal part) of the way $(x(t), y(t), z(t))$ is from P to Q. We will often use this standard formula to parametrize line segments.

Mathematica's `ParametricPlot3D` command will produce a plot of a curve in space, given its parametric equations, just as `ParametricPlot` produces a plot of a plane curve from its parametric equations.

A surface is also very naturally described by parametric equations, only now because the surface is two-dimensional it requires two parameters to specify the location of a typical point on the surface. The parametric equations will have the form $\begin{cases} x = x(s,t) \\ y = y(s,t), \\ z = z(s,t) \end{cases}$ where the point (s, t) ranges over an appropriate "parameter region" in the st plane. We've already discussed one example of a

139

surface described parametrically: the plane through P parallel to two non-parallel vectors **v** and **w** is traced out by the point $\mathbf{x}(s, t) = \mathbf{P} + s\mathbf{v} + t\mathbf{w}$ as the parameters s and t independently range over all possible real values.In the examples below we will show how to find parametric equations for some other surfaces. We can then use *Mathematica*'s `ParametricPlot3D` command to visualize the surfaces. The `ParametricPlot3D` command requires that the parameter region be a rectangle with sides parallel to the s and t axes, so we will always find parametric equations meeting this requirement.

Example 1. Find parametric equations for the graph of $z = x^2 y$ over the rectangle $0 \leq x \leq 1$, $-1 \leq y \leq 2$, and use the equations to plot the surface.
Solution:

We use the x and y coordinates as our two parameters: $\begin{cases} x = s \\ y = t \\ z = s^2 t \end{cases}$, so the

parameter region is simply the rectangle $0 \leq s \leq 1$, $-1 \leq t \leq 2$. Note that *the same method can be used to find parametric equations for the graph of any function $f(x, y)$ over a rectangle with sides parallel to the coordinate axes.*
The command
`ParametricPlot3D[{s, t, s^2 t}, {s,0,1}, {t,-1,2}]`
will then produce a perspective view of this surface, much like that produced by the command `Plot3D[x^2 y, {x,0,1}, {y,-1,2}]`. The `Plot3D` command is much faster, but is limited to plotting graphs of functions.

Example 2. Find parametric equations for the circular cylinder $x^2 + y^2 = 9$, for $0 \leq z \leq 4$, and display this cylinder.
Solution:

This surface is not the graph of a function $z = f(x, y)$, so the method used in the preceding example will not apply. Here the idea is to let one parameter measure location on the circle $x^2 + y^2 = 9$ in the xy plane, and let the other parameter be the z-coordinate. Let $z = s$, and parametrize the circle as usual: $\begin{cases} x = 3\cos t \\ y = 3\sin t \end{cases}$, $0 \leq t \leq 2\pi$. The portion of the cylinder we want consists of all points (x, y, z) whose x and y coordinates satisfy $x^2 + y^2 = 9$ and whose z-coordinate is between 0 and 4. The equations

$\begin{cases} x = 3\cos t \\ y = 3\sin t \\ z = s \end{cases}$ generate precisely these points, as (s, t) ranges over $\begin{cases} 0 \leq s \leq 4 \\ 0 \leq t \leq 2\pi \end{cases}$.

Consequently, the command

`ParametricPlot3D[{3Cos[t], 3Sin[t], s}, {s,0,4}, {t,0,2Pi}]`

will produce a perspective view of this cylinder. Note that *the same method can be used to find parametric equations for the generalized cylinder obtained by translating __any__ parametrized curve in the xy-plane, parallel to the z-axis.*

Options for ParametricPlot3D

The options available for the `ParametricPlot3D` command are much like those of `Plot3D`, discussed earlier. In particular, to have equal scales on the three axes, the option `BoxRatios -> Automatic` should be used. Because there are two independent parameters for a surface, the value of the `PlotPoints` option is a list of two positive integers. For example if we had used the option `PlotPoints->{5,15}` in Example 2 above, then 5 equally spaced values for the first parameter s would have been selected from the s-interval [0, 4], namely 0, 1, 2, 3, 4. and these would be combined with 15 equally spaced values in the interval $0 \leq t \leq 2\pi$, to give $5 \times 15 = 75$ sample points (s, t), which would then be substituted into the parametric equations to find the coordinates of the corner points of the quadrilateral faces forming the mesh surface. Thus rather than having equal numbers of sample values selected from the s and t intervals, we can adjust these numbers independently if we wish. (If a value is not specified for `PlotPoints`, *Mathematica* uses the default value {15, 15}) In the plot of the cylinder above, *the surface is not curved in the vertical direction*, so we could use only two sample s-values:

`PlotPoints -> {2,15}` would yield only $2 \times 15 = 30$ sample points, greatly speeding up the production of the plot without any loss in accuracy. If we wanted a smoother approximation to the cylinder, we could increase the number of sample points for the polar angle parameter t, say to 30. Thus the command

`ParametricPlot3D[{3Cos[t],3Sin[t],s}, {s,0,4},`
`    {t,0,2Pi}, BoxRatios->Automatic, PlotPoints->{2,30}]`

would give a much better plot of the cylinder, in less time, than the basic command used above. (The $1 \times 29 = 29$-faced mesh surface is much more quickly displayed than the default $14 \times 14 = 196$-faced mesh surface, and it is a better approximation of the cylindrical surface.)

Example 3. The cylinder $x^2 + y^2 = 9$ discussed above in Example 2 is an example of a surface of revolution — the cylinder is traced out when we revolve the vertical line from $(3, 0, 0)$ to $(3, 0, 4)$ about the z-axis. Generalize the reasoning in Example 2 to find parametric equations for the surface traced out when an arbitrary curve $\begin{cases} x = x(s) \\ z = z(s) \end{cases}$, $a \le s \le b$ is revolved about the z-axis. Assume that $x(s) \ge 0$ for all s in $[a, b]$, for simplicity.

Solution:

When the point $(x(s), 0, z(s))$ is turned about the z-axis through an angle t, the coordinates of the resulting point are $(x(s)\cos t, x(s)\sin t, z(s))$. As t runs from 0 to 2π the point $(x(s)\cos t, x(s)\sin t, z(s))$ traces a circle of radius $x(s)$ at height $z(s)$, centered on the z-axis. So <u>parametric equations for the surface of revolution traced out by revolving the entire curve</u> $\begin{cases} x = x(s) \\ z = z(s) \end{cases}$, $a \le s \le b$ <u>around the</u>

<u>z-axis are</u> $\begin{cases} x = x(s)\cos t \\ y = x(s)\sin t \\ z = z(s) \end{cases}$, $\begin{cases} a \le s \le b \\ 0 \le t \le 2\pi \end{cases}$. This formula will prove useful in the exercises below. The circular cylinder in Example 2 is a special case, where the line from $(3, 0, 0)$ to $(3, 0, 4)$ is parametrized by $\begin{cases} x = 3 \\ z = s \end{cases}$, $0 \le s \le 3$.

Example 4. Make a perspective picture of the solid bounded by the cylinder $x^2 + y^2 = 9$, the xy plane and the plane $3x + 2y + z = 12$. Then find the volume of this solid.

Solution:

The plane $z = 12 - 3x - 2y$ is the graph of a function, so we can plot it using `ParametricPlot3D` as described in Example 1. (We can use just two sample points for each parameter, as discussed above, since the plane is not curved.) And we've just seen how to draw the cylinder in Example 2. Note that the highest point on the graph of $z = 12 - 3x - 2y$ for $-3 \le x \le 3$ and $-3 \le y \le 3$ would be 27, which occurs when x and y both equal -3. So we display the portion of our cylinder with $0 \le z \le 27$. Thus the *Mathematica* commands

```
plane = ParametricPlot3D[{s, t, 12-3s-2t}, {s,-3,3},
    {t,-3,3},PlotPoints->{2,2}, DisplayFunction->Identity]
```

```
cylinder  =  ParametricPlot3D[{3Cos[t],3Sin[t],s},
    {s,0,27},  {t,0,2Pi},  PlotPoints->{2,30},
    DisplayFunction->Identity]

Show[plane,cylinder,BoxRatios->Automatic,ViewPoint->{2,1,1},
    DisplayFunction->$DisplayFunction]
```

will display the plane and cylinder together. From this crude sketch we can easily set up a double integral to evaluate the volume of our solid, the portion of space under the graph of $z = 12 - 3x - 2y$ and above the disk R: $x^2 + y^2 = 9$ in the xy plane, $\iint\limits_{R} 12 - 3x - 2y \, dA$. This equals the iterated integral

$$\int_{-3}^{3} \int_{-\sqrt{9-x^2}}^{\sqrt{9-x^2}} 12 - 3x - 2y \, dy dx, \text{ and the command}$$

`Integrate[12-3x-2y,{x,-3,3},{y,-Sqrt[9-x^2],Sqrt[9-x^2]}]`. then gives the result 108 Pi.

It would be nice to be able to display just the part of the plane $z = 12 - 3x - 2y$ which lies above the disk $x^2 + y^2 = 9$, and just the part of the cylinder that lies below the plane, so that the solid of interest would be shown without any excess portions of the plane and cylinder. This can be done — all that is required is a little more care in finding the parametric equations. First, note that if we substitute the expressions $\begin{cases} x = 3\cos t \\ y = 3\sin t \end{cases}$ into the equation of the plane, we get the z-coordinates of the points on the plane which lie directly above points on the circle $x^2 + y^2 = 9$. Thus parametric equations for this curve of intersection, an ellipse, are $\begin{cases} x = 3\cos t \\ y = 3\sin t \\ z = 12 - 9\cos t - 6\sin t \end{cases}$, $0 \le t \le 2\pi$.

We can now modify these equations to get parametric equations for the portion of the plane inside the cylinder and for the portion of the cylinder which lies below the plane.

For the portion of the plane inside the cylinder just replace the radius 3 in the parametric equations of the ellipse by a variable radius r, and let r range from 0 to 3. Every point inside the ellipse lies on the plane $z = 12 - 3x - 2y$ directly above some point $\begin{cases} x = r\cos t \\ y = r\sin t \end{cases}$, where $\begin{cases} 0 \le r \le 3 \\ 0 \le t \le 2\pi \end{cases}$.

143

So the equations $\begin{cases} x = r\cos t \\ y = r\sin t \\ z = 12 - 3r\cos t - 2r\sin t \end{cases}$, $\begin{matrix} 0 \le r \le 3 \\ 0 \le t \le 2\pi \end{matrix}$ parametrize the portion of

the plane $z = 12 - 3x - 2y$ inside the cylinder $x^2 + y^2 = 9$.

Similarly, for every t in $[0, 2\pi]$ the point P = $(3\cos t, 3\sin t, 0)$ lies on the circle $x^2 + y^2 = 9$ directly below the point Q = $(3\cos t, 3\sin t, 12 - 9\cos t - 6\sin t)$, and the line segment joining these two points lies on the cylinder $x^2 + y^2 = 9$. Using the standard parametrization for the line segment from P to Q, we get the

parametric equations $\begin{cases} x = 3\cos t \\ y = 3\sin t \\ z = s(12 - 9\cos t - 6\sin t) \end{cases}$, $\begin{matrix} 0 \le s \le 1 \\ 0 \le t \le 2\pi \end{matrix}$ for the portion of the

cylinder lying between the plane $z = 0$ and the plane $z = 12 - 3x - 2y$.

The *Mathematica* commands

```
top = ParametricPlot3D[{r Cos[t],r Sin[t],
    12 - 3r Cos[t] - 2r Sin[t]}, {r,0,3}, {t,0,2Pi},
    PlotPoints->{2,30}, DisplayFunction -> Identity]

sides = ParametricPlot3D[{3 Cos[t], 3 Sin[t],
    s (12 - 9Cos[t] - 6Sin[t])}, {s,0,1}, {t,0,2Pi},
    PlotPoints->{2,30}, DisplayFunction -> Identity]

Show[top, sides, BoxRatios->Automatic,
    AxesLabel->{"x","y","z"},ViewPoint->{3,1,1},
    DisplayFunction->$DisplayFunction]
```

will produce a clear picture of our solid. Changing the `ViewPoint` would allow us to view the solid from any perspective. (We've left off the bottom disk, of course, but if desired it could also be included.)

Remark. Examples 3 and 4 contain a concentrated dose of geometry! Study them carefully to make sure you understand the geometric interpretation of the parameters and their ranges. These examples typify the mixture of geometric and algebraic reasoning required in finding parametric equations for surfaces. With the recent advent of powerful graphics systems like *Mathematica*, techniques for parametrizing curves and surfaces have suddenly assumed much greater practical importance. Not only are parametric equations needed for displaying curves or surfaces with the `ParametricPlot3D` command — we will also need to find parametric equations in the next chapter, when we will

144

be integrating functions and vector fields over curves and surfaces. Thus although the following exercises may require extra thought, the effort will pay handsome dividends.

Exercises

1. Find parametric equations for the cone obtained by revolving the line segment from (0, 0, 4) to (2, 0, 0) around the z-axis. Use a **ParametricPlot3D** command to display this cone.

2. Find parametric equations for the sphere of radius 5 centered at (0, 0, 3), and display the sphere. (Think of the sphere as the surface traced out by revolving a semicircle in the xz plane about the z-axis. Start by parametrizing the appropriate semicircle.)

3. Find parametric equations for the surfaces of the solid bounded above by the graph of $z = 9 - x^2 - y^2$ and below by the graph of $z = x^2 + y^2 + 1$. Display this solid, and find its volume. (*Hint:* What is the curve of intersection of the two surfaces?)

4. (§17.2 #41) Find parametric equations for the surfaces of the wedge cut from the elliptic cylinder $4x^2 + y^2 = 9$ by the planes $z = 0$ and $z = y + 3$. Display the wedge, using a suitable **ViewPoint**, and calculate its volume using a double integral.

5. (§17.2 #42) Find parametric equations for the surfaces of the solid in the first octant bounded above by the graph of $z = 9 - x^2$, below by $z = 0$, and laterally by the plane $y = 0$ and the parabolic cylinder $y^2 = 3x$. Display the solid and calculate its volume. (First, show that the curve of intersection of the two parabolic cylinders can be parametrized by $\begin{cases} x = t^2/3 \\ y = t \\ z = 9 - t^4/9 \end{cases}$, $0 \le t \le 3$.)

6. Here is a geometric problem involving Lagrange multipliers in which the equations to be solved are too complex for *Mathematica*'s **Solve** command. We can use a plot to crudely estimate the solution, and then use the **FindRoot** command to locate the solution to any desired accuracy. The problem is to find the highest point on the curve of intersection of the plane $2x + y + z = 1$ with the torus obtained by revolving the circle $y^2 + (z - 2)^2 = 1$ around the y-axis.

(i) Find parametric equations $\begin{cases} x = x(s,t) \\ y = y(s,t) \\ z = z(s,t) \end{cases}$ for the torus, a surface of revolution.

Note the geometric meaning of the two surface parameters s and t.
Use **ParametricPlot3D** commands and a **Show** command to plot the torus together with the portion of the plane with $-3 \le x \le 3$ and $-3 \le z \le 3$. (Parametrize the plane as the graph of the function $y = 1 - 2x - z$ over this square.) From the plot, estimate the coordinates of the highest point on the intersection of the plane with the torus. Estimate the values of the parameters s and t at this highest point.

(ii) Our problem is to maximize $z(s, t)$, subject to the condition that $2x(s, t) + y(s, t) + z(s, t) = 1$. Find the Lagrange equations for this constrained maximization problem. Enter this system of equations into *Mathematica*. (Use L as the *Mathematica* name for the Lagrange multiplier λ.)

(iii) The Lagrange equations involve trigonometric functions of s and t, so *Mathematica*'s **Solve** command is unable to solve them. To find a solution of this system of equations using the **FindRoot** command, we need a fairly good numerical estimate of the desired solution. You've estimated values for s and t from the plot in part (i) above, but what about an estimate of λ? Substitute the estimated values for s and t at the highest point into the Lagrange equations to get a system of three equations, each involving only the Lagrange multiplier λ. These equations will be inconsistent, since presumably the estimated values of s and t were not exactly those at the maximum. Solve each of these equations for λ and use the average of the three values as the starting value for λ in the **FindRoot** command. Thus you'll execute the command
FindRoot[equations,{s,s0},{t,t0},{L,L0}]
if **equations** is the name given to the set of Lagrange equations, and **s0,t0, L0** are the estimated values of s, t and λ corresponding to the highest point. Check your result by substituting the values of s and t into the parametric equations for the torus, to verify that the coordinates of the corresponding point do appear to be those of the highest point on the curve of intersection of the torus with the plane.

§4 Surface area

The text [1] describes how to express the surface area of the portion of the graph of a function $f(x, y)$ that lies above or below a rectangle R in the xy plane:
$$\iint_R \sqrt{f_x(x,y)^2 + f_y(x,y)^2 + 1}\ dA.$$ This formula is a special case of a more general formula for surface area, which applies to surfaces described by parametric equations. Since we have invested considerable effort in learning to parametrize surfaces, it seems a shame not to make use of this expertise when discussing surface area. The area formula for parametric surfaces is best understood in vector notation, so we first digress a moment to review this notation.

If the parametric equations for a surface σ are $\begin{cases} x = x(s,t) \\ y = y(s,t), \ (s,t) \in R, \text{ we will} \\ z = z(s,t) \end{cases}$

group the three components to form a vector $\mathbf{x}(s, t) = \ <x(s, t), y(s, t), z(s, t)>$. Thus as the point (s, t) ranges over the parameter region R, the point $\mathbf{x}(s, t)$ ranges over the surface σ. The vector of partial derivatives $\mathbf{x}_s(s,t)$ is the tangent vector at $\mathbf{x}(s, t)$ to the curve on σ through the point $\mathbf{x}(s, t)$, found by holding t fixed, in the parametric equations, and letting s increase. Similarly $\mathbf{x}_t(s,t)$ is the tangent vector to the curve through $\mathbf{x}(s, t)$ found by holding s fixed and letting t increase. We assume that except possibly at a finite number of points (s, t) the vectors $\mathbf{x}_s(s,t)$ and $\mathbf{x}_t(s,t)$ are nonzero and non-parallel. Thus, except at a finite number of points of the surface σ the cross product $\mathbf{x}_s(s,t) \times \mathbf{x}_t(s,t)$ is a nonzero vector normal to the tangent plane of σ at $\mathbf{x}(s, t)$. The length $|\mathbf{x}_s(s,t) \times \mathbf{x}_t(s,t)|$ of this vector is equal to the area of the parallelogram with the two tangent vectors as adjacent sides, and the direction of $\mathbf{x}_s(s,t) \times \mathbf{x}_t(s,t)$ is such that the ordered triple of vectors $(\mathbf{x}_s(s,t),\ \mathbf{x}_t(s,t),\ \mathbf{x}_s(s,t) \times \mathbf{x}_t(s,t))$ is right-handed.

By an argument similar to the one in the text [1] which led to the surface area formula $\iint_R \sqrt{f_x(x,y)^2 + f_y(x,y)^2 + 1}\ dA$ for graphs of functions, one can

justify the more general surface area formula $\boxed{\text{Area of } \sigma = \iint_R \|\mathbf{x}_s(s,t) \times \mathbf{x}_t(s,t)\|\ dA}$.

We shall not give this justification, for lack of space, but we can easily show that for graphs of functions the two formulas are equivalent. Recall that the standard way to parametrize the portion of the graph of $f(x, y)$ which lies above or below a rectangle R is to take x and y as the parameters: $\begin{cases} x = s \\ y = t \\ z = f(s,t) \end{cases}$, $(s,t) \in R$.

Then $\mathbf{x}(s, t) = < s, t, f(s, t)>$, so $\mathbf{x}_s(s,t) = < 1, 0, f_x(s,t)>$ and $\mathbf{x}_t(s,t) = < 0, 1, f_y(s,t)>$, from which it is easy to calculate $\|\mathbf{x}_s(s,t) \times \mathbf{x}_t(s,t)\| = \sqrt{f_x(s,t)^2 + f_y(s,t)^2 + 1}$.

Remark: An easy way to remember the area formulas for parametric surfaces and graphs of functions is to note their similarity to the arc-length formula for a parametric curve $\mathbf{x} = \mathbf{x}(t)$, $a \le t \le b$: $L = \int_a^b \|\mathbf{x}'(t)\| \, dt$, and the special case for the graph of a function $y = f(x)$, $a \le x \le b$: $L = \int_a^b \sqrt{f'(x)^2 + 1} \, dx$.

Example 1 Find the surface area of the cylinder $x^2 + y^2 = 9$, $0 \le z \le 4$.
Solution:

We've seen in Example 2 of the preceding section that parametric equations for this cylinder are $\begin{cases} x = 3\cos t \\ y = 3\sin t \\ z = s \end{cases}$, $\begin{cases} 0 \le s \le 4 \\ 0 \le t \le 2\pi \end{cases}$.

Thus $\mathbf{x}(s, t) = < 3\cos t, 3\sin t, s >$. We could easily calculate the tangent vectors and their cross product by hand in this example, but this is a good opportunity to let *Mathematica* do some of the work. (We first load the LinearAlgebra package, which contains the command for computing cross products of vectors, into the computer's memory.)

```
Needs["LinearAlgebra`CrossProduct`"]
X[s_,t_]  =  {3Cos[t],3Sin[t],s}
Xs  =  D[X[s,t],s]
Xt  =  D[X[s,t],t]
normal  =  Cross[Xs,Xt]
```
The final output is {-3 Cos[t], -3 Sin[t], 0}. Thus $\|\mathbf{x}_s(s,t) \times \mathbf{x}_t(s,t)\| = 3$, and the surface area is $\int_0^4 \int_0^{2\pi} 3 \, dt \, ds = 24\pi$. Replacing the radius 3 of the cylinder by a variable r and replacing the height by h, our calculation would give the general formula for the surface area of a cylinder: $S = 2\pi r h$.

Because the vector $\mathbf{x}_s(s,t) \times \mathbf{x}_t(s,t)$ is a nonzero normal vector at the point $\mathbf{x}(s, t)$, we can easily write equations for the tangent plane and the normal line at this point. In vector form an equation for the tangent plane is $(\mathbf{x}_s(s,t) \times \mathbf{x}_t(s,t)) \cdot (\mathbf{x} - \mathbf{x}(s,t)) = 0$, and an equation for the normal line is $\mathbf{x}(s,t) = \mathbf{x} = \mathbf{x}(s,t) + u(\mathbf{x}_s(s,t) \times \mathbf{x}_t(s,t))$, $-\infty \leq u \leq \infty$. Expressed another way, if $\mathbf{x}(s, t) = (x_0, y_0, z_0)$ is any point on the surface and $\mathbf{x}_s(s,t) \times \mathbf{x}_t(s,t) = (a, b, c)$ is the corresponding normal vector, then the equation of the tangent plane at this point is $a(x - x_0) + b(y - y_0) + (z - z_0) = 0$ and parametric equations for the normal

line are $\begin{cases} x = x_0 + au \\ y = y_0 + bu, \quad -\infty \leq u \leq \infty. \\ z = z_0 + cu \end{cases}$

Example 2 Find an equation for the tangent plane to the cylinder $x^2 + y^2 = 9$, $0 \leq z \leq 4$ in Example 1 above, at the point $(2, \sqrt{5}, 3)$.

Solution:

Using the parametrization of the surface from Example 1, we first find the values of the parameters s and t corresponding to the point $(2, \sqrt{5}, 3)$. To do this, we solve the equations $\begin{cases} 3\cos t = 2 \\ 3\sin t = \sqrt{5}, \\ s = 3 \end{cases}$ to get $s = 3$, $t = \arccos(\frac{2}{3})$ (which is the same as $\arcsin(\frac{\sqrt{5}}{3})$). Using the calculation of $\mathbf{x}_s(s,t) \times \mathbf{x}_t(s,t)$ in Example 1, the *Mathematica* command

```
ReplaceAll[normal,{s -> 3, t -> N[ArcCos[2/3]} ]
```

will produce the normal vector {–2., –2.23607, 0} at the point $(2, \sqrt{5}, 3)$. Knowing the point and the normal vector, we can write the equation of the tangent plane: $2(x - 2) + 2.23607(y - \sqrt{5}) = 0$. The absence of z in this equation tells us that the tangent plane is parallel to the z-axis, as is the cylinder itself.

Exercises

1. Plot the surface with parametric equations $\begin{cases} x = s\cos t \\ y = s\sin t, \\ z = t \end{cases}$ $\begin{matrix} 0 \leq s \leq 2 \\ 0 \leq t \leq 4\pi \end{matrix}$.

(This surface, called a helicoid, is not easily expressible as a level surface $f(x, y, z) = c$. It is an example of a *minimal surface*, i.e., it is the surface of minimal area whose boundary is the section of the z-axis and the helix formed by setting $s = 0$ and $s = 2$.) Compute the surface area of the helicoid.

2. (§17.4 #10)

i) Express the surface area of the portion of the sphere $x^2 + y^2 + z^2 = 8$ that is inside the cone $z = \sqrt{x^2 + y^2}$ as a double integral, and evaluate this integral.

ii) Parametrize the sphere and the cone, and display the two surfaces together in a single picture.

iii) Find parametric equations for the curve of intersection of the sphere and cone, and display this curve.

iv) Find parametric equations for the portion of the sphere that lies inside the cone, and display this surface, whose area was found in part (i).

3. (§17.4 #4)

i) Express the surface area of the portion of the cone $z = \sqrt{x^2 + y^2}$ that lies inside the cylinder $x^2 + y^2 = 2x$ as a double integral, and evaluate this integral.

ii) Parametrize the cone and the cylinder, and display the two surfaces together in a single picture.

iii) Find parametric equations for the curve of intersection of the cone and cylinder, and display this curve.

iv) Find parametric equations for the portion of the cone that lies inside the cylinder, and display this surface, whose area was found in part (i).

4. Find parametric equations for the torus (doughnut-shaped surface) obtained by revolving the circle $(x-5)^2 + z^2 = 9$ around the z-axis. Display this torus and find its surface area.

5. Find parametric equations for the normal line to the torus in Exercise 4, at the point $\left(\dfrac{13\sqrt{3}}{4}, \dfrac{13}{4}, \dfrac{3\sqrt{3}}{2} \right)$. Plot the torus together with the segment of the normal line of length 2, centered at this point.

150

§5 Triple integrals

The greatest difficulty most students have with triple integrals is visualizing the solid region of integration. By learning to find parametric equations for surfaces one can overcome this obstacle. The boundary surfaces of the region of integration can be displayed by using the **ParametricPlot3D** command to plot each surface, then the surfaces can be combined in a single picture using a **Show** command. By choosing a suitable **ViewPoint**, if necessary, the projection of the surface onto any of the coordinate planes can be visualized, making easy the evaluation of the triple integral by means of interated integrals.

Example Use *Mathematica* to display the solid region G bounded by the paraboloid $z = 4x^2 + y^2$ and the parabolic cylinder $z = 4 - 3y^2$. Evaluate $\iiint\limits_{G} yz\,dV$.

Solution:

First we parametrize the two surfaces separately and combine them in a single plot. Both surfaces are parametrized as graphs of functions:

```
paraboloid = ParametricPlot3D[{s,t,4s^2 + t^2},
    {s,-1.2,1.2},  {t,-1,1},PlotPoints->{8,10}]
cylinder = ParametricPlot3D[{s,t,4 - 3t^2},
    {s,-1,1},  {t,-1,1},PlotPoints->{2,10}]
Show[paraboloid,cylinder,AxesLabel->{"x","y","z"}]
```

(Since the cylinder is not curved in the direction of the x-axis, we use only two sample s-values to keep the number of faces of the mesh surface as small as possible.) If necessary, one could re-parametrize just the portions of the two surfaces which form the boundaries of the region G. This is not necessary if our goal is to evaluate a triple integral over G, because from the output of the **Show** command it is clear that the projection of G onto the yz plane is the region R between the parabola $z = 4 - 3y^2$ and the parabola $z = y^2$ obtained by setting $x = 0$ in the equation of the paraboloid. One can easily plot R:

```
Plot[{4 - 3y^2, y^2},  {y,-2,2},  AxesLabel->{"y","z"}]
```

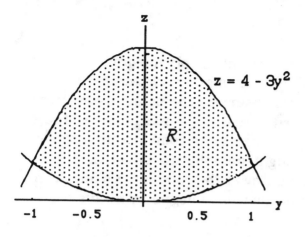

The surface of G nearest R is half of the parabolic cylinder, which we can express as a function of (y, z) over R: $x = \dfrac{-\sqrt{z-y^2}}{2}$. Similarly the other half of the parabolic cylinder forms the opposite surface of G: $x = \dfrac{\sqrt{z-y^2}}{2}$. Thus

$$\iiint\limits_G yz\, dV = \iint\limits_R \left(\int_{-\frac{\sqrt{z-y^2}}{2}}^{\frac{\sqrt{z-y^2}}{2}} yz\, dx \right) dA, \text{ and since } \int_{-\frac{\sqrt{z-y^2}}{2}}^{\frac{\sqrt{z-y^2}}{2}} yz\, dx = yz\sqrt{z-y^2}, \text{ we must}$$

evaluate $\displaystyle\iint\limits_R yz\sqrt{z-y^2}\, dA$. From the sketch of R this equals $\displaystyle\int_{-1}^{1}\int_{y^2}^{4-3y^2} yz\sqrt{z-y^2}\, dz\, dy$. This we can turn over to *Mathematica*, using the command

```
Integrate[y z Sqrt[z - y^2],{y,-1,1},{z,y^2,4-3y^2}]
```

which yields the result 0. In retrospect we can see that this is correct by evaluating $\displaystyle\iint\limits_R yz\sqrt{z-y^2}\, dA$ by integrating with respect to y first: since $yz\sqrt{z-y^2}$ is an odd function of y then, because the region R is symmetric about the z-axis, for each z the integral with respect to y will be 0.

Mathematica note: *Mathematica*'s **Integrate** command can evaluate iterated triple integrals, provided that it is able to find the required antiderivatives.

Example Evaluate $\int_0^3 \int_{1-x}^{1+x^2} \int_{2-3x-y}^{x^2+y^2+4} x^2 y + yz \, dz \, dy \, dx$, using *Mathematica.*

Solution:

The command

```
Integrate[x^2 y + y z,  {x,0,3},{y,1-x,1+x^2},
    {z,2-3x-y,x^2 + y^2 + 4}]
```

produces the exact value, $\dfrac{3450528009}{80080}$.

Exercises

1. (§17.5 #10) Use *Mathematica* to display the solid region G bounded by the plane $z = y$, the xy-plane, and the parabolic cylinder $y = 1-x^2$. Then evaluate the triple integral $\iiint_G y \, dV$.

2. (§17.5 #18) Use *Mathematica* to display the solid region G common to the cylinders $x^2 + y^2 = 1$ and $x^2 + z^2 = 1$. Then use a triple integral to find the volume of G.

3. (§17.5 #26) The iterated integral $\int_0^3 \int_0^{\sqrt{9-z^2}} \int_0^{\sqrt{9-y^2-z^2}} f(x,y,z) \, dx \, dy \, dz$ is equivalent to a triple integral over a solid region G. Use *Mathematica* to display the region G. Then express the triple integral as an iterated integral in which the z-integration is performed first, the y-integration second and the x-integration last.

4. The coordinates $(\bar{x}, \bar{y}, \bar{z})$ of the center of mass of a solid with the shape of a region G, with density $f(x, y, z)$ at any point (x, y, z) in G, are

$$\bar{x} = \frac{\iiint_G x f(x,y,z) \, dV}{m}, \quad \bar{y} = \frac{\iiint_G y f(x,y,z) \, dV}{m}, \quad \bar{z} = \frac{\iiint_G z f(x,y,z) \, dV}{m}, \quad \text{where}$$

$m = \iiint_G f(x,y,z) \, dV$ is the mass of the solid.

Use these formulas to find the center of mass of the solid hemisphere bounded by the xy-plane and the surface $z = \sqrt{r^2 - x^2 - y^2}$, if the density of the solid is constant.

§6 Triple integrals in cylindrical and spherical coordinates

Changing to spherical or cylindrical coordinates may greatly simplify the calculations needed to evaluate certain integrals. Parametric equations for the surfaces bounding the region of integration, required if we want to plot these surfaces using the **ParametricPlot3D** command, must still be expressed in cartesian coordinates. To find these parametric equations we use the equations

$$\begin{cases} x = r\cos\theta \\ y = r\sin\theta \end{cases} \quad \text{and} \quad \begin{cases} x = \rho\sin(\phi)\cos(\theta) \\ y = \rho\sin(\phi)\sin(\theta), \\ z = \rho\cos(\phi) \end{cases}$$ and other relations between the cartesian

coordinates of a point and its cylindrical or spherical coordinates.

Example Use *Mathematica* to display the solid G bounded above by the sphere $\rho = 4$ and below by the cone $\phi = \pi/3$. Find the centroid of G.

Solution:

The surfaces bounding this solid are so familiar that we can make a fairly accurate sketch of G by hand. But we'll use this as an opportunity to practice parametrizing surfaces which have been described by equations in spherical coordinates.

Parametric equations for the sphere are easy — we just take the zenith angle ϕ and the azimuth angle θ as parameters: $\begin{cases} x = 4\sin(s)\cos(t) \\ y = 4\sin(s)\sin(t), \\ z = 4\cos(s) \end{cases} \begin{cases} 0 \le s \le \pi \\ 0 \le t \le 2\pi \end{cases}$.

(Because *Mathematica* does not allow us to use Greek letters as names of variables, we've replaced ϕ by s and θ by t.)

Using the relation $\tan\phi = \dfrac{\sqrt{x^2 + y^2}}{z}$ and the fact $\tan(\pi/3) = \sqrt{3}$, the equation of the

cone can be written $\sqrt{3} = \dfrac{\sqrt{x^2 + y^2}}{z}$, or $z = \dfrac{\sqrt{x^2 + y^2}}{\sqrt{3}}$. Eliminating z from the

cartesian equations of the cone and sphere: $x^2 + y^2 = 3z^2$ and $x^2 + y^2 + z^2 = 16$ shows that the cone intersects the sphere above the circle $x^2 + y^2 = 12$ in the xy-plane. (This could also be shown by a rough sketch.) Thus, taking the cylindrical coordinates r and θ as parameters, equations for the portion of this

cone inside the sphere are $\begin{cases} x = r\cos t \\ y = r\sin t, \\ z = r/\sqrt{3} \end{cases} \begin{cases} 0 \le r \le \sqrt{12} = 2\sqrt{3} \\ 0 \le t \le 2\pi \end{cases}$.

So the commands

```
sphere  =  ParametricPlot3D[{4Sin[s]  Cos[t],4Sin[s]Sin[t],
          4Cos[s]},{s,0,Pi/3},{t,0,2Pi},PlotPoints->{10,10}]

cone  =  ParametricPlot3D[{r  Cos[t],r  Sin[t],r/Sqrt[3]},
          {r,0,2Sqrt[3]},{t,0,2Pi},PlotPoints->{2,10}]

Show[sphere,cone,ViewPoint->{3,   0.1,   0},
      BoxRatios->Automatic]
```

will display the surface of the solid G. By the symmetry of G about the z-axis, its centroid lies on this axis. The height of the centroid is given by the formula

$$\bar{z} = \frac{\iiint\limits_G z\,dV}{\iiint\limits_G 1\,dV},$$ so we must work out these two integrals. Here is where the

descriptions of the surfaces in spherical coordinates come in handy:

$$\iiint\limits_G 1\,dV \;=\; \int_0^{2\pi}\int_0^{\pi/3}\int_0^4 \rho\sin\phi\,d\rho\,d\phi\,d\theta \;=\; 16\pi\int_0^{\pi/3}\sin\phi\,d\phi \;=\; 8\pi, \text{ and}$$

$$\iiint\limits_G z\,dV \;=\; \int_0^{2\pi}\int_0^{\pi/3}\int_0^4 \rho^2\cos\phi\sin\phi\,d\rho\,d\phi\,d\theta \;=\; \frac{128\pi}{3}\int_0^{\pi/3}\cos\phi\sin\phi\,d\phi \;=\; 16\pi.$$

We conclude that the centroid is $(0, 0, 2)$.

Exercises

1. For each of the following problems, first find parametric equations for each
    surface bounding the solid, and use the **ParametricPlot3D** command to
    plot the surface. (If you can, parametrize just the portion of the surface
    which actually bounds the solid, since this will give the best picture of the
    solid.) Then combine the plots of the individual surfaces into a single plot,
    using a **Show** command. Finally, express the quantity requested as a triple
    integral and evaluate it, using cylindrical or spherical coordinates.

i) (§17.7 #9) Find the volume of the solid in the first octant inside both the
    sphere $x^2 + y^2 + z^2 = 16$ and the cylinder $x^2 + y^2 = 4x$.

ii) (§17.7 #11) Find the volume of the solid in the first octant bounded by the
    sphere $\rho = 2$, the coordinate planes, and the cones $\phi = \pi/6$ and $\phi = \pi/3$.

iii) (§17.7 #24) Find the centroid of the solid bounded above by the paraboloid $z = x^2 + y^2$, below by the plane $z = 0$, and laterally by the cylinder $(x-1)^2 + y^2 = 1$.

iv) Find the volume of the surface of revolution $\rho = 1 + \cos\phi$.

2. (Supplementary Exercises for Chapter 17, #28)
(a) Change to cylindrical coordinates and then evaluate

$$\int_{-2}^{2} \int_{-\sqrt{4-x^2}}^{\sqrt{4-x^2}} \int_{(x^2+y^2)^2}^{16} x^2\, dz\, dy\, dx.$$

(b) Change to spherical coordinates and then evaluate

$$\int_{0}^{1} \int_{0}^{\sqrt{1-x^2}} \int_{0}^{\sqrt{1-x^2-y^2}} \frac{1}{1+x^2+y^2+z^2}\, dz\, dy\, dx$$

3. The iterated integral $I = \int_{0}^{1} \int_{x}^{\sqrt{2-x^2}} \int_{\sqrt{x^2+y^2}}^{\sqrt{4-x^2-y^2}} x^2 + y^2 + z^2\, dz\, dy\, dx$ is equivalent to a triple integral over a solid region D.

i) Parametrize the boundary surfaces of the solid D and use **ParametricPlot3D** commands and a **Show** command to display D.

ii) Express I as an iterated integral in cylindrical coordinates.

iii) Express I as an iterated integral in spherical coordinates.

Chapter 15 Integrals over Curves and Surfaces

§ 1 Line integrals

This section summarizes the different types of integrals over curves in $\Re^2$ or $\Re^3$. Vector notation is used to describe these so-called "line integrals", to emphasize the similarity between $\Re^2$ and $\Re^3$.

Integrals of scalar functions over curves

Let C be a smooth curve in $\Re^3$, parametrized by $\mathbf{x}(t) = <x(t), y(t), z(t)>$, $a \le t \le b$. If $f(x, y, z)$ is a function on $\Re^3$ which is defined along the curve C, the integral of f over C, denoted $\int_C f(x,y,z)\,ds$ is defined to be $\int_a^b f(\mathbf{x}(t)) \|\mathbf{x}'(t)\|\,dt$. A basic theorem asserts that the value of this integral is independent of the choice of parametrization of the curve C. If the function f is interpreted as a charge (or mass) density on a "wire" in the shape of C, then $\int_C f(x,y,z)\,ds$ represents the total charge (or mass) of the wire.

Example 1 Find the total charge on a wire bent in the shape of the circle $x^2 + z^2 = 9$ in the xz plane, if the charge density along the wire is given by

$$f(x, y, z) = \frac{k}{x^2 + y^2 + z^2}.$$

Solution:

Start by parametrizing the curve: $\begin{cases} x = 3\sin t \\ y = 0 \\ z = 3\cos t \end{cases}$, $0 \le t \le 2\pi$. Using

Mathematica, we define this as a vector-valued function,

```
X[t_]  =  {3Sin[t],  0,  3Cos[t]}
```

then define the density function as a scalar-valued function of a vector

```
f[{x_,y_,z_}]  =  k/(x^2  +  y^2  +  z^2)
```

and integrate: (We use the formula $\|\mathbf{v}\| = \sqrt{\mathbf{v} \cdot \mathbf{v}}$ to compute $\|\mathbf{x}'(t)\|$.)

```
Integrate[f[X[t]]  Sqrt[  X'[t].X'[t]  ],{t,0,2Pi}]
```

We can check the result $\frac{2\mathrm{P}ik}{3}$: the charge density has the constant value $k/9$ everywhere on our circle, so the value of the integral is this constant times $\int_0^{2\pi} \|\mathbf{x}'(t)\|\,dt$, and this last integral is the circumference of the circle, 6π.

We conclude that the total charge on the wire is indeed $2\pi k/3$.

Mathematica Note: We've used the capital letter **x** to denote vector-valued function parametrizing the curve. This leaves the lower case **x** available as a name for the x-component function $x(t)$ of the vector-valued function **x[t]**. Also note the use of the list brackets in the definition **f[{x_,y_,z_}]**. We want **f** to be a function of a vector {x, y, z}, so that the composite **f[X[t]]** will be defined

Integrating the tangential component of a vector field along a curve

The unit tangent vector to the oriented curve C is $\mathbf{T}(t) = \dfrac{\mathbf{x}'(t)}{\|\mathbf{x}'(t)\|}$. If $\mathbf{F}(x,y,z)$ is a vector field defined on C, the integral of the tangential component of $\mathbf{F}$ along C, i.e., the integral of the scalar function $\mathbf{F}\cdot\mathbf{T}$, which is naturally denoted

$\displaystyle\int_C \mathbf{F}\cdot\mathbf{T}\,ds$, is $\displaystyle\int_a^b \left(\mathbf{F}(\mathbf{x}(t))\cdot\dfrac{\mathbf{x}'(t)}{\|\mathbf{x}'(t)\|}\right)\|\mathbf{x}'(t)\|\,dt$, or simply $\displaystyle\int_a^b \mathbf{F}(\mathbf{x}(t))\cdot\mathbf{x}'(t)\,dt$. If we write out the components of the vector field, say $\mathbf{F}(x,y,z) = <f_1(x,y,z), f_2(x,y,z), f_3(x,y,z)>$, then

$$\int_a^b \mathbf{F}(\mathbf{x}(t))\cdot\mathbf{x}'(t)\,dt = \int_a^b f_1(x(t),y(t),z(t))\,x'(t) + f_2(x(t),y(t),z(t))\,y'(t) + f_3(x(t),y(t),z(t))\,z'(t)\,dt.$$

(We can consider plane curves and vector fields on $\mathfrak{R}^2$ as a special case of this, in which $z(t) \equiv 0$, $f_3(x,y,z) \equiv 0$ and z does not occur in f_1 and f_2, i. e., $\mathbf{x}(t) = <x(t), y(t)>$ and $\mathbf{F}(x,y) = <f_1(x,y), f_2(x,y)>$.)

We record the definition for reference:

Definition. The integral of the tangential component of a vector field $\mathbf{F}$ along a curve C, denoted $\displaystyle\int_C \mathbf{F}\cdot\mathbf{T}\,ds$, is defined to be $\displaystyle\int_a^b \mathbf{F}(\mathbf{x}(t))\cdot\mathbf{x}'(t)\,dt$, or in expanded form

$$\int_a^b f_1(x(t),y(t),z(t))\,x'(t) + f_2(x(t),y(t),z(t))\,y'(t) + f_3(x(t),y(t),z(t))\,z'(t)\,dt.$$

Other notations for this integral are $\displaystyle\int_C \mathbf{F}(x,y,z)\cdot d\mathbf{r}$, which is based on the notation $\mathbf{r}(t)$ rather than $\mathbf{x}(t)$ for the vector representation of the curve C, and the "differential form" notation $\displaystyle\int_C f_1(x,y,z)dx + f_2(x,y,z)dy + f_3(x,y,z)dz$ which is common in textbooks in the physical sciences and engineering.

If we interpret the vector $\mathbf{F}(x,y,z)$ as the force acting on a particle located at the point (x, y, z), then the integral $\int_C \mathbf{F} \cdot \mathbf{T}\, ds$ represents the <u>work</u> done by the force field $\mathbf{F}$ during a motion of the particle along the curve C. When the orientation of C is reversed, the unit tangent vector $\mathbf{T}$ changes sign so the sign of the integral $\int_C \mathbf{F} \cdot \mathbf{T}\, ds$ changes too.

Evaluation of integrals $\int_C \mathbf{F} \cdot \mathbf{T}\, ds$ is straightforward using *Mathematica*, since the program handles vector operations very simply.

Example 2 Evaluate $\int_C \mathbf{F} \cdot \mathbf{T} ds$ if C is the circle $x^2 + y^2 = 4$, traversed counterclockwise, and $\mathbf{F}$ is the plane vector field $\mathbf{F}(x,y) = <x^2 y - 3y, -2xy>$.
Solution:

We begin by parametrizing the circle in the standard way: $\begin{cases} x = 2\cos t \\ y = 2\sin t \end{cases}$,
$0 \le t \le 2\pi$. Define the vector function $\mathbf{x}(t)$ and the vector field $\mathbf{F}(x,y)$ as *Mathematica* functions:

```
X[t_] = {2Cos[t], 2Sin[t]}
F[{x_,y_}] = {x^2 y - 3y, -2x y}.
```
Then the command

`Integrate[F[X[t]].X'[t], {t,0,2Pi}]` instructs *Mathematica* to attempt to evaluate the integral symbolically, using its integral tables. (If the symbolic integration should fail, we could replace the `Integrate` command by `NIntegrate`, to get a numerical approximation.) Using this example as a template, the only difficulty one faces in evaluating line integrals of vector fields is in finding parametric equations for the path of integration.

Remarks

1. User-defined *Mathematica* variables generally should have names that begin with a lower case letter, to avoid conflicts with names which have already been reserved for built-in *Mathematica* constants or commands. (For example the letter **E** is reserved for Euler's constant $e \approx 2.71828$, **N** denotes *Mathematica*'s "numerically evaluate" command, and **C** is reserved for the constants of integration which occur in solving differential equations.) However, to make our *Mathematica* notation parallel the standard mathematical notation as closely as possible, we have used the capital characters **F** and **X** to denote the vector field and the vector of functions which defines the parametrization of C.

2. Note the use of the list brackets in the definition `F[{x_,y_}]`. We want `F` to be a function of a vector {x, y}, so that the composite `F[X[t]]` will be defined.

Example 3 Evaluate the line integral $\int_C x^2 z\, dx + 2(y+z)\, dy - xy^2 z\, dz$, where C is the triangular path from (1, 0, 0) to (1, 3, 1) to (4, 2, 5) and back to (1, 0, 0).

Solution:

We define the vector field as before:

```
F[{x_, y_, z_}] = {x^2 z,  2(y+z),  -x y^2 z},
```

and parametrize each of the three sides of the triangular path, being careful to orient the line segments correctly.

```
X1[t_] = {1,0,0} + t ({1,3,1} - {1,0,0})
X2[t_] = {1,3,1} + t ({4,2,5} - {1,3,1})
X3[t_] = {4,2,5} + t ({1,0,0} - {4,2,5}).
```

Then just integrate the tangential component of **F** over each of the sides, and add the results:

```
side1 = Integrate[F[X1[t]].X1'[t],  {t,0,1}]
side2 = Integrate[F[X2[t]].X2'[t],  {t,0,1}]
side3 = Integrate[F[X3[t]].X3'[t],  {t,0,1}]
sum = side1 + side2 + side3
```

Integrating the normal component of a plane vector field along a curve

On a smooth plane curve the tangent vector $\mathbf{x}'(t) = \langle x'(t), y'(t) \rangle$ is never zero. The vector $\mathbf{n}(t) = \langle y'(t), -x'(t) \rangle$ results from turning $\mathbf{x}'(t)$ clockwise through 90°, so it has the same length as $\mathbf{x}'(t)$, but is perpendicular to $\mathbf{x}'(t)$. Thus the vector function $\mathbf{N}(t) = \dfrac{\mathbf{n}(t)}{\|\mathbf{x}'(t)\|}$ gives the unit normal vector at $\mathbf{x}(t)$ obtained by turning the unit tangent vector $\mathbf{T}(t)$ clockwise through 90°. If $\mathbf{F}(x, y) = \langle f_1(x,y), f_2(x,y) \rangle$ is a vector field on $\Re^2$ which is defined along the plane curve C, the integral of the normal component of **F** over C, denoted $\int_C \mathbf{F} \cdot \mathbf{N}\, ds$ is $\int_a^b (\mathbf{F}(\mathbf{x}(t)) \cdot \mathbf{N}(t)) \|\mathbf{x}'(t)\|\, dt = \int_a^b \mathbf{F}(\mathbf{x}(t)) \cdot \mathbf{n}(t)\, dt$. Other ways to express this integral are the differential form $\int_C f_1(x,y)\, dy - f_2(x,y)\, dx$, or in full detail $\int_a^b f_1(x(t),y(t))\, y'(t) - f_2(x(t),y(t))\, x'(t)\, dt$. The value of this integral depends on the orientation of C but otherwise is independent of the choice of parametrization of the curve.

If the vector field $\mathbf{F}(x, y)$ is interpreted as the velocity of flow of a fluid at the point (x, y), then the integral of the tangential component of $\mathbf{F}$ along C, in suitable units, measures the average flow of fluid along the curve, and the integral of the normal component of $\mathbf{F}$ along C measures the average flow of the fluid from left to right across C. This fluid flow interpretation is often indicated by referring to $\int_C \mathbf{F} \cdot \mathbf{T}\, ds$ as the "flow of $\mathbf{F}$ along C", and calling $\int_C \mathbf{F} \cdot \mathbf{N}\, ds$ the "flux of $\mathbf{F}$ across C". We will see that the three-dimensional analogues of these ideas are important later in this chapter.

Example 4 If C is the circle $x^2 + y^2 = 4$ and $\mathbf{F}(x, y) = <x^2 y - 3y, -2xy>$, calculate the flux of $\mathbf{F}$ across C.

Solution:

As usual, we begin by parametrizing the path of integration:

```
X[t_]  =  {2Cos[t],  2Sin[t]}.
```
The command `n[t_]  =  {X'[t][[2]],  -X'[t][[1]]}` computes the normal vector $\mathbf{n}(t)$ to our circle. (Recall that if $\mathbf{v}$ denotes any vector (list) in *Mathematica*, then `v[[1]]` denotes its first entry, `v[[2]]` its second entry, etc.)

Then we just define the vector field as before, and integrate:

```
F[{x_,  y_}]  =  {x^2 y  -  3y,  -2x y}
Integrate[F[X[t]].n[t],   {t,0,2Pi}]
```

Summary

1. The integral of a scalar function along a curve:

$$\int_C f(x,y,z)\, ds = \int_a^b f(\mathbf{x}(t))\, \|\mathbf{x}'(t)\|\, dt$$

2. The integral of the tangential component of a vector field along a curve:

$$\int_C \mathbf{F} \cdot \mathbf{T}\, ds = \int_a^b \mathbf{F}(\mathbf{x}(t)) \cdot \mathbf{x}'(t)\, dt = \int_C f_1(x,y,z)dx + f_2(x,y,z)dy + f_3(x,y,z)dz =$$

$$\int_a^b f_1(x(t),y(t),z(t))\, x'(t) + f_2(x(t),y(t),z(t))\, y'(t) + f_3(x(t),y(t),z(t))\, z'(t)\, dt$$

3. The integral of the normal component of a vector field along a <u>plane</u> curve:

$$\int_C \mathbf{F} \cdot \mathbf{N}\, ds = \int_a^b \mathbf{F}(\mathbf{x}(t)) \cdot \mathbf{n}(t)\, dt = \int_C f_1(x,y)\, dy - f_2(x,y)\, dx =$$

$$\int_a^b f_1(x(t),y(t))\, y'(t) - f_2(x(t),y(t))\, x'(t)\, dt$$

Exercises

1. Find the total mass of a wire in the shape of the helix $\begin{cases} x = 3\cos(\pi t) \\ y = 3\sin(\pi t) \\ z = 2t \end{cases}$, $0 \le t \le 5$, if the density of the wire is given by the formula $f(x, y, z) = \dfrac{1}{z+1}$.

2. The coordinates $(\bar{x}, \bar{y}, \bar{z})$ of the center of mass of a wire bent in the shape of a curve C, with density $f(x, y, z)$ at any point (x, y, z) on C, are

$$\bar{x} = \frac{\int_C x f(x,y,z)\,ds}{m}, \quad \bar{y} = \frac{\int_C y f(x,y,z)\,ds}{m}, \quad \bar{z} = \frac{\int_C z f(x,y,z)\,ds}{m}, \quad \text{where } m = \int_C f(x,y,z)\,ds \text{ is the}$$

mass of the wire.

   Use these formulas to find the center of mass of the helical wire in Exercise 1 above. Plot the helix, indicating the location of the center of mass of the wire in the plot. (Note: The command

```
point  =  Graphics3D[{RGBColor[1,0,0],PointSize[.05],
                  Point[{xbar,ybar,zbar}]}]
```

will create a *Graphics3D* object consisting of a red point with coordinates {xbar, ybar, zbar}, which can be displayed together with other *Graphics3D* objects, such as the output of a **ParametricPlot3D** command, by using a **Show** command. The diameter of the point will be 5% of the width of the final plot.)

3. Let $\mathbf{F}(x, y) = <\dfrac{x}{x^2+y^2}, \dfrac{y}{x^2+y^2}>$, if $(x, y) \ne (0,0)$, and let C_1 be the curve consisting of the upper semicircular arc from $(4, 0)$ to $(-4, 0)$, followed by the line segment from $(-4\ 0)$ to $(-1, 0)$, followed by the upper semicircular arc from $(-1, 0)$ to $(1, 0)$, and finally the line segment from $(1, 0)$ to $(4, 0)$.

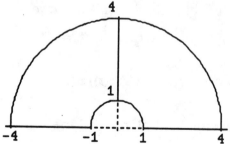

i) Calculate the flow $\int_{C_1} \mathbf{F} \cdot \mathbf{T}\, ds$ of $\mathbf{F}$ along C_1 and the flux $\int_{C_1} \mathbf{F} \cdot \mathbf{N}\, ds$ of $\mathbf{F}$ across this

   path.

ii) Replace the radius of the inner semicircle by an arbitrary value r > 0, and let the resulting path be denoted C_r. Find the limit of $\int_{C_r} \mathbf{F} \cdot \mathbf{T} \, ds$ and of $\int_{C_r} \mathbf{F} \cdot \mathbf{N} \, ds$ as $r \to 0$. (This limit may be thought of as defining the improper integrals of the tangential and normal components of $\mathbf{F}$ along a path through the singular point of this vector field at the origin.)

iii) Repeat parts (i) and (ii), but with the path changed so that the inner semicircle is the lower half of a circle, as in the sketch below. (The results show that using different paths to define improper line integrals may give different answers.)

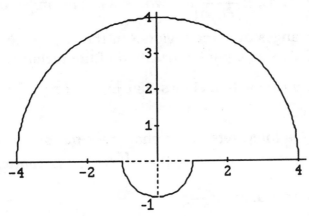

§ 2 Surface integrals

This section describes integrals of scalar functions and vector fields over surfaces in $\mathfrak{R}^3$. Vector notation and parametric equations are employed to emphasize the parallels with integrals over curves.

Integrals of scalar functions over surfaces

Let σ be a smooth surface in $\mathfrak{R}^3$, parametrized by $\mathbf{x}(s,t) = \langle x(s,t), y(s,t), z(s,t) \rangle$, where (s,t) ranges over a region R in the st-plane. If $f(x, y, z)$ is a function on $\mathfrak{R}^3$ which is defined along the surface σ, the integral of f over σ, denoted $\iint_\sigma f(x,y,z) \, dS$, is defined to be $\iint_R f(\mathbf{x}(s,t)) \|\mathbf{x}_s(s,t) \times \mathbf{x}_t(s,t)\| \, dS$. A basic theorem asserts that the value of this integral is independent of the choice of parametrization of the surface σ.

Remarks

1. Note the similarity between the formulas defining the integral of a function $f(x, y, z)$ over a surface and over a curve:

$$\begin{cases} \iint_\sigma f(x,y,z)\,dS = \iint_R f(\mathbf{x}(s,t))\|\mathbf{x}_s(s,t) \times \mathbf{x}_t(s,t)\|\,dS \\ \int_C f(x,y,z)\,ds = \int_a^b f(\mathbf{x}(t))\|\mathbf{x}'(t)\|\,dt \end{cases}$$

2. The formula $\iint_\sigma f(x,y,z)\,dS = \iint_R f(\mathbf{x}(s,t))\|\mathbf{x}_s(s,t) \times \mathbf{x}_t(s,t)\|\,dS$ can be viewed as a generalization of the formula for changing variables in a double integral:

$\iint_S f(x,y)\,dA = \iint_R f(x(s,t),y(s,t))\left|\dfrac{\partial(x,y)}{\partial(s,t)}\right|dA$, where as (s, t) ranges over the region R the point $(x(s,t), y(s,t))$ ranges over the region S in the xy-plane. We simply think of $\mathbf{x}(s, t) = \,< x(s,t), y(s,t), 0 >$ as a parametrization of the region S, viewed as a surface in $\mathfrak{R}^3$. Then an easy calculation shows that $\|\mathbf{x}_s(s,t) \times \mathbf{x}_t(s,t)\| = \left|\dfrac{\partial(x,y)}{\partial(s,t)}\right|$.

3. If the function f is interpreted as a charge (or mass) density on a curved sheet or "lamina" in the shape of σ, then $\iint_\sigma f(x,y,z)\,dS$ represents the total charge (or mass) of the lamina.

Example 1 Find the total charge on a lamina in the shape of the cylinder $x^2 + y^2 = 9$, $0 \le z \le 4$, if the charge density along the lamina is given by $f(x, y, z) = \dfrac{k}{x^2 + y^2 + z^2}$.

Solution:

Start by parametrizing the surface: $\begin{cases} x = 3\cos t \\ y = 3\sin t, \\ z = s \end{cases} \quad \begin{array}{c} 0 \le s \le 4 \\ 0 \le t \le 2\pi \end{array}$. Using

Mathematica, we define this as a vector-valued function:

```
X[s_,t_]  =  {3Cos[t],  3Sin[t],  s}
```

Because this function involves trigonometric functions, we will load the Trigonometry package so that the **TrigReduce** command will be available

```
Needs["Algebra`Trigonometry`"]
```

Using the identity $\|\mathbf{u}\times\mathbf{v}\|=\left(\|\mathbf{u}\|^2\|\mathbf{v}\|^2-(\mathbf{u}\cdot\mathbf{v})^2\right)^{1/2}$, we can compute the length of the cross product of two vectors using dot products alone, without computing the cross product itself. (Since the factor $\|\mathbf{x}_s(s,t)\times\mathbf{x}_t(s,t)\|$ can be interpreted as the factor by which the mapping $\mathbf{x}$: $(s,\,t)\to\mathbf{x}(s,t)$ from R onto σ magnifies oriented area at $(s,\,t)$, we will give the name mag(s,t) to the *Mathematica* function corresponding to $\|\mathbf{x}_s(s,t)\times\mathbf{x}_t(s,t)\|$.)

```
Xs = D[X[s, t], s]
Xt = D[X[s, t], t]
mag[s_,t_] = Sqrt[TrigReduce[(Xs.Xs)  (Xt.Xt)  -  (Xs.Xt)^2]  ]
```

Then define the charge density as a scalar-valued function of a vector, and integrate:

```
f[{x_,y_,z_}]  =  k/(x^2 + y^2 + z^2)
Integrate[f[X[s,t]]  mag[s,t],{s,0,4},{t,0,2Pi}]
```

The resulting output, 2 Pi k ArcTan$[\frac{4}{3}]$, gives the total charge on the lamina.

Integrals of vector fields over oriented surfaces

An oriented surface is intuitively a surface with two distinct sides. We can express this mathematically by saying that there are two *continuous unit normal vector fields* on such a surface — one in which all the vectors point toward one side of the surface, and the negative of this in which all vectors point toward the opposite side of the surface.

The vectors $\mathbf{x}_s(s,t)$ and $\mathbf{x}_t(s,t)$ are tangent to the surface σ at the point $\mathbf{x}(s,\,t)$ so their cross product is a normal vector at this point. On a smooth surface this normal vector $\mathbf{n}(s,\,t)=\mathbf{x}_s(s,t)\times\mathbf{x}_t(s,t)$ is never zero, so as $(s,\,t)$ ranges over the parameter region R, the vectors $\mathbf{n}(s,\,t)$ form a continuous nowhere zero vector field on the surface σ. Thus the unit normal vectors $\mathbf{N}(s,\,t)=\dfrac{\mathbf{x}_s(s,t)\times\mathbf{x}_t(s,t)}{\|\mathbf{x}_s(s,t)\times\mathbf{x}_t(s,t)\|}$ all point to one side of σ — they provide an *orientation* of the surface. The two possible orientations of σ (continuous unit normal vector fields on σ) are therefore expressible parametrically: If $\mathbf{N}(x,y,z)$ is any orientation of σ then $\mathbf{N}[\mathbf{x}(s,\,t)]=\pm\mathbf{N}(s,\,t)$. Since $\mathbf{x}_t(s,t)\times\mathbf{x}_s(s,t)=-\mathbf{x}_s(s,t)\times\mathbf{x}_t(s,t)$, we can choose the order of these tangent vectors in forming their cross product to produce either of the two orientations of σ.

Now suppose $\mathbf{F}(x, y, z)$ is a vector field defined on σ, and $\mathbf{N}$ is an orientation of the surface σ. And suppose $\mathbf{N}(\mathbf{x}(s, t)) = \mathbf{N}(s, t)$, i.e., $\dfrac{\mathbf{x}_s(s,t) \times \mathbf{x}_t(s,t)}{\|\mathbf{x}_s(s,t) \times \mathbf{x}_t(s,t)\|}$. The integral over σ of the component of $\mathbf{F}$ along $\mathbf{N}$, denoted $\iint\limits_{\sigma} \mathbf{F} \cdot \mathbf{N}\, dS$, is defined to be the integral of the scalar function $\mathbf{F} \cdot \mathbf{N}$ over the surface. That is,

$$\iint\limits_{\sigma} \mathbf{F} \cdot \mathbf{N}\, dS = \iint\limits_{R} \mathbf{F}(\mathbf{x}(s,t)) \cdot (\mathbf{x}_s(s,t) \times \mathbf{x}_t(s,t))\, dA = \iint\limits_{R} \mathbf{F}(\mathbf{x}(s,t)) \cdot \mathbf{n}(s,t)\, dA.$$

If $\mathbf{F}(x, y, z)$ is interpreted as the velocity of a fluid flow at (x, y, z), then in suitable units $\iint\limits_{\sigma} \mathbf{F} \cdot \mathbf{N}\, dS$ represents the average amount of fluid which crosses the surface per unit time. Thus this integral is often called the "flux" of $\mathbf{F}$ across σ in the direction of $\mathbf{N}$.

Example 2 If $\mathbf{F}(x, y, z) = \langle x, y, z^2 \rangle$, and σ is the sphere $x^2 + y^2 + z^2 = 4$, find the outward flux of $\mathbf{F}$ across σ.
Solution:

As always, begin by parametrizing the surface. Taking the zenith and azimuth angles as parameters, we have $\begin{cases} x = 2\sin\phi\cos\theta \\ y = 2\sin\phi\sin\theta \,, \\ z = 2\cos\phi \end{cases} \begin{array}{l} 0 \le \phi \le \pi \\ 0 \le \theta \le 2\pi \end{array}$. Using *Mathematica* we define the curve and the vector field:

```
X[s_,t_]={2Sin[s] Cos[t], 2Sin[s] Sin[t], 2Cos[s]}
F[{x_,y_,z_}] = {x, y, z^2}
```

Next, we compute and simplify $\mathbf{n}(s, t) = \mathbf{x}_s(s,t) \times \mathbf{x}_t(s,t)$:

```
Needs["Algebra`Trigonometry`"]
Needs["LinearAlgebra`CrossProduct`"]
n[s_,t_] = TrigReduce[ Cross[ D[X[s,t],s], D[X[s,t],t] ] ]
```

The result is clearly an outward normal vector on the sphere, rather than an inward normal. (One could also see this by visualizing the tangent vectors $\mathbf{x}_\phi$ and $\mathbf{x}_\theta$ pointing in the directions of increasing zenith and azimuth angles respectively, and using the right-hand rule to see that the cross product in the order $\mathbf{x}_\phi \times \mathbf{x}_\theta$ would be an outward normal.)

Finally, compute the dot product $\mathbf{F}(\mathbf{x}(s,t)) \cdot (\mathbf{x}_s(s,t) \times \mathbf{x}_t(s,t))$ and integrate

```
integrand  =   TrigReduce[F[X[s,t]].n[s,t]]
```

```
Integrate[integrand,   {s,0,Pi},   {t,0,2Pi}]
```

The output, $\dfrac{64\ Pi}{3}$, is the outward flux of $\mathbf{F}$ across the sphere.

Summary

1. The integral of a scalar function over a surface:
$$\iint_\sigma f(x,y,z)\,dS = \iint_R f(\mathbf{x}(s,t)) \|\mathbf{x}_s(s,t) \times \mathbf{x}_t(s,t)\|\,dS$$

2. The flux of a vector field across an oriented surface:
$$\iint_\sigma \mathbf{F} \cdot \mathbf{N}\,dS = \iint_R \mathbf{F}(\mathbf{x}(s,t)) \cdot \mathbf{n}(s,t)\,dA,$$ where $\mathbf{n}(s,\ t)$ is either $\mathbf{x}_s(s,t) \times \mathbf{x}_t(s,t)$ or
$\mathbf{x}_t(s,t) \times \mathbf{x}_s(s,t)$, the order of the two tangent vectors $\mathbf{x}_s(s,t)$ and $\mathbf{x}_t(s,t)$ in the cross product being chosen so that $\mathbf{n}(s,\ t)$ will have the same direction as the given orientation $\mathbf{N}(\mathbf{x}(s,\ t))$.

Exercises

(§18.5 #4) Calculate the indicated unit normal to the surface at the specified point.

(a) The unit normal to the surface $y^2 = x$ at $(1,1,2)$ that points toward the xz-plane.

(b) the unit normal to the hyperbolic paraboloid $y = z^2 - x^2$ at $(1,3,2)$ that points toward the yz-plane.

(c) the unit normal to the cone $x^2 = y^2 + z^2$ at $(\sqrt{2},-1,-1)$ that points away from the xy-plane.

2. Find the total mass of a lamina with the shape of the portion of the paraboloid $z = 16 - x^2 - y^2$ above the xy plane, if the density of the lamina at any point $(x,\ y,\ z)$ is kz, for some constant k.

3. The coordinates $(\bar{x}, \bar{y}, \bar{z})$ of the center of mass of a lamina curved in the shape of a surface σ, with density $f(x, y, z)$ at any point (x, y, z) on σ, are

$$\bar{x} = \frac{\iint\limits_{\sigma} x\, f(x,y,z)\, dS}{m}, \qquad \bar{y} = \frac{\iint\limits_{\sigma} y\, f(x,y,z)\, dS}{m}, \qquad \bar{z} = \frac{\iint\limits_{\sigma} z\, f(x,y,z)\, dS}{m},$$

where $m = \iint\limits_{\sigma} f(x,y,z)\, dS$ is the mass of the lamina.

Use these formulas to find the center of mass of the lamina in Exercise 2 above.

4. (§18.5 #8) Evaluate $\iint\limits_{\sigma} \mathbf{F} \cdot \mathbf{N}\, dS$, if $\mathbf{F}(x, y, z) = \left\langle x^2, \ yx, \ zx \right\rangle$ and σ is the portion of the plane $6x + 3y + 2z = 6$ in the first octant, where $\mathbf{N}$ is the unit normal vector field on σ with positive z-component.

5. Evaluate $\iint\limits_{\sigma} \mathbf{F} \cdot \mathbf{N}\, dS$, if $\mathbf{F}(x, y, z) = \langle x, y, z \rangle$, σ is the surface of the closed cylinder $x^2 + y^2 \le 1$, $0 \le z \le 2$, and $\mathbf{N}$ is the outward unit normal on σ.

6. Evaluate $\iint\limits_{\sigma} \mathbf{F} \cdot \mathbf{N}\, dS$, if $\mathbf{F}(x, y, z) = \left\langle x^2, \ y^2, \ z^2 \right\rangle$, σ is the sphere $x^2 + y^2 + z^2 = 9$, and $\mathbf{N}$ is the outward unit normal on σ.

Index